WARGAMING
AIRBORNE OPERATIONS

90 mm statuette, in pewter, of a British parachutist. Designed by Charles Stadden and marketed by Hamilton Marriott (London).

WARGAMING

AIRBORNE OPERATIONS

DONALD FEATHERSTONE

SOUTH BRUNSWICK AND NEW YORK:
A. S. BARNES AND COMPANY

FIRST AMERICAN EDITION

1979

Library of Congress Catalogue Card Number: 78-53669

A. S. Barnes and Co., Inc.
Cranbury, New Jersey 08512

ISBN 0-498-02250-1

Printed in the United States of America

CONTENTS

'There has been no single performance by any unit that has more greatly inspired me or more excited my admiration than the nine day action by the 1st British Parachute Division between 17th and 25th September.'

Dwight Eisenhower, General U.S. Army
Allied Commander Europe

THE EARLY HISTORY OF AIRBORNE FORCES

As armies returned to traditional soldiering in the years that followed World War One, the paratroop concept vanished in all nations save Russia. Germany and Russia were the great industrial nations defeated in World War One and this perhaps lessened their respect for the old methods which had failed them. Hence they were the first to experiment with airborne forces, demonstrating that air-landed infantry with ancillary support troops were an integral element of modern airborne operations.

Heavily affected by the purges of the 1930s, the Soviet airborne forces took only marginal roles in the wars of 1939 and 1941/45, being few in number and relatively limited in scale. In the Finnish War of 1939 two separate parachute brigades were badly scattered in their drops and failed to gain their objectives; heavy pilot and transport losses in the early months of the German invasion were a further set-back. Like the German Fallschirmjager after Crete, the Soviet airborne soldiers fought as conventional infantry throughout World War Two, or supported guerillas in small groups, carried out reconnaissance, sabotage or intelligence missions.

In 1935 the Germans, after extensive observation of training in Russia, formed an experimental airborne staff commanded by Brigadier-General Kurt Student, an officer in the Reichswehr and an infantry and airforce veteran of World War One. Soon the Luftwaffe and the army formed Fallschirmjager battalions who demonstrated the eminent practicability of this unique method of troop deployment on the battlefield, proving that the concept was sufficiently valid to be accepted as a valuable military tactic. The part played by airborne warfare in the events of May 1940 (although formed in 1939 the Germans 7th Air Division was not used in Poland), convinced military leadership that something new and worthwhile had arrived.

After Crete, the German Fallschirmjager divisions were committed as infantry on the Eastern Front where they won a formidable reputation as tough fighters, causing the German High Command to appreciate this availability of high quality infantry without being particularly concerned at their use in a role other than that for which they were trained.

At the beginning of World War Two the strategic employment of armour completely changed the concepts of warfare as carried over from World War One. At the end of World War Two, all the major powers had experience of the airborne method, both as attackers and as attacked, and the victors possessed in various measures means of combating such operations and some understanding of the potential of the new method. It is not inconceivable that, at the beginning of a future war, large airborne forces will play a role similar to that of German armour in 1940.

Despite being misunderstood and mis-used, both Allied and Axis airborne forces captured the public imagination more than any other aspect of ground warfare since the decline of cavalry. Testing a soldier's courage and manhood in the fiery crucible of battle, they created and perpetuated elitism, encouraged by extra pay, distinctive colourful insignia and uniforms.

THE FORMATION OF ALLIED AIRBORNE FORCES

There were great obstacles to the formation of airborne forces – a concept that needed men to jump, plus pilots and planes to fly them into battle; all with specialised training. Also, it required the support and endorsement of senior commanders and politicians – the far-seeing Prime Minister flung himself into the fray. On 27 May 1941, Churchill wrote to General Lord Ismay:

'. . . in the light of what is happening in Crete and may soon be happening in Cyprus and in Syria . . . we ought to have an airborne division on the German model with any improvements which might suggest themselves from experience. We ought also to have a number of carrier aircraft.'

On 22 June 1940, Churchill sent a minute to his Chiefs of Staff: 'We ought to have a corps of at least 5,000 parachute troops. I hear something is being done already to form such a corps but only, I believe on a very small scale. Advantage must be taken of the summer to train these forces, who can none the less play their part meanwhile as shock troops in home defence.'

Service planners were sceptical of Churchill's idea of taking the war to the enemy on land in Europe and believed parachute units to be logistically unsound. Nevertheless, in July 1940 the Central Landing Establishment was formed at Ringway near Manchester, in three sub-units: the Parachute Sqn., the Glider Sqn. and Technical Development Unit with R.A.F. and army staffs recruited to organise parachute training and evolve the logistics of airborne assault.

In May 1941 it was proposed to Winston Churchill that the future composition of airborne forces should be two parachute brigades, one in the U.K. and one in the Middle East, and a glider force sufficient to lift 10,000 men and associated equipment; ten medium bomber squadrons were to be converted to paratroop dropping and glider towing role. Thus the small airborne cadre expanded into brigades and later divisions that were to play major parts in future operations.

Brigadier R. N. Gale, a ruddy-faced moustached World War One veteran who wore riding breeches and polished field boots but possessed all the qualities of a first-class British Senior Officer, was appointed to command 1st Parachute Brigade of three battalions, 11th SAS became 1st Parachute Battalion. In September 1941 volunteers between 22 and 32 years of age were called for from all infantry battalions in the United Kingdom; not more than ten men were to be taken from any one unit – the standards of fitness were very high. Lieut. Colonel Brown took over the 1st Battalion, the 2nd was raised under Lieut. Colonel Flavell and the 3rd under Lieut. Colonel Lathbury. In mid-1942 a new recruitment policy drafted trained infantry battalions, from famous County regiments, who brought their own regimental traditions to benefit the fledgling formation of which they had become a part. Men of these regiments who failed the selection course were replaced by volunteers from other sources. The first converted battalions were 7th Queen's Own Cameron Highlanders who brought their own pipe band with them, and 10th Royal Welch Fusiliers, becoming 5th (Scottish) and 6th (Royal Welch) Parachute Battalions. From 1st Parachute Brigade was transferred the 4th Battalion (later the Wessex) and on 17 July 1942 Brigadier Down formed 2nd Parachute Brigade of 4th, 5th and 6th Battalions.

On 1 August 1942 the War Office approved the formation of the Parachute Regiment to serve in conjunction with the Glider Pilot Regiment as part of the Army Air Corps. These airborne units were welded together as a fighting machine by Brigadier F. A. M. Browning, then commanding 24th Guards Brigade, who was selected to be 'Commander Paratroops and Airborne Troops' with the rank of Major-General. A colourful personality, gifted with the style and looks of a film star and the acme of sartorial elegance, 'Boy' Browning had all the qualities of superb leadership and it was he who chose the maroon beret as a distinctive headgear of the Airborne Division, and the emblem of Bellerophon mounted on the winged horse Pegasus to adorn the arm of all airborne soldiers.

In late 1942 Brigadier Hackett's 4th Parachute Brigade was formed in Egypt and Northern Palestine. It consisted of the 10th (Sussex) Parachute Battalion (formerly 2nd Battalion Royal Sussex Regiment); the 11th from volunteers serving in units in the Middle East, and the 151st Parachute Battalion was the original 156th (British) Parachute Battalion transferred from India.

In May 1943 6th Airborne Division was formed. Its 3rd Parachute Brigade consisted of 7th, 8th and 9th Parachute Battalions formed of the following battalions impressed into the Parachute Regiment: 7th (Light Infantry) Bn. (from 10th Bn. Somerset Light Infantry); 8th (Midland) Bn. (from 13th Bn. Royal Warwickshire Regiment); and 9th (Eastern and Home Counties) Bn. (from 10th Bn. Essex Regiment). On 1 June, the War Office re-named 10th Bn. Green Howards (2nd/4th Bn. South Lancashire Regiment) as 12th (Yorkshire) and 13th (Lancashire) Parachute Battalions of Brigadier Poett's 5th Brigade to make up the 6th Airborne Division with Hill's 3rd Brigade. The Division was commanded by Major-General Richard Gale.

In 1942 a British Division numbered 757 officers and 16,764 men in two infantry brigades, each of three rifle battalions, and a tank brigade. It was supported by three field regiments of artillery, an anti-tank regiment, and a light anti-aircraft regiment. Other support elements included Royal Engineers, Divisional Signals and a Reconnaissance Regiment; service elements were an Army Service Corps for transport, Royal Army Medical Corps, Royal Army Ordnance Corps and Provost plus other minor units. H.Q. Staff was assisted by Intelligence, Field Security and Defence sections.

In the British services the term 'Airborne' covers parachute and glider-borne troops, raised, organised, trained and equipped for those respective roles. An airborne division was formed of many elements: the Parachute Regiment, the glider pilots and glider-borne infantry, plus essential support and service units. Its establishment followed the same overall pattern, the tank brigade being replaced by the Air Landing Brigade, and with a drastic cut in ancillary services to achieve air mobility. This reduced the strength of the division by about a third as compared with the conventional infantry division.

The divisional artillery of an airborne division was made up of three batteries of the Air Landing Light Regiment, plus two anti-tank batteries and a light anti-aircraft battery. The 3.7 Howitzer originally used by the 1st Landing Light Regiment was discarded in favour of the American 75 mm Pack Howitzer, actually a mule-borne mountain gun that, when pneumatised, proved ideal for transportation in a glider. An airborne battery had six guns and was made up of sub-sections consisting of a gun, two jeeps and three trailers to carry ammunition and stores. A typical gunload for a Horsa glider was a gun, a jeep, a trailer and one or two motorcycles and six passengers.

1st Airborne Division had its own artillery, the 1st Airborne Light Regiment R.A.; General Gale's 6th Airborne Division's artillery was provided by the 53rd (Worcestershire Yeomanry) Air Landing Regiment R.A. equipped with American 75 mm Howitzers.

Each Parachute Brigade had a squadron of a Reconnaissance Regiment allocated to it, equipped with seven-ton Tetrarch tanks, jeeps and motorcycles; Hamilcar gliders were devised for carrying the Tetrarch into action. At Arnhem, the 1st Airborne Division's reconnaissance facilities were reduced to a single squadron with armoured jeeps. Parachute brigades had their own squadron of Royal Engineers and a company was with each Air Landing Brigade; the Royal Signals maintained Divisional Wireless and external links with artillery, air-support, etc. Standard British Army radio sets, from the 18/68 to 76 sets were used; most communication was by the 38 set (0.5 watt output) operating on voice up to a maximum of four miles. A parachute field ambulance, consisting of a H.Q., two surgical teams and four sections of nine RAMC officers and 100 other ranks, was attached to each parachute brigade; the air-landing field ambulance had an increased establishment. Specially designed airborne medical and surgical equipment was dropped in arms containers or carried in gliders.

A Parachute Battalion numbered 600/800 all ranks, with three or sometimes four companies each totalling five officers and 120 men plus battalion H.Q. and H.Q. Company; the company was divided into H.Q. and three 36-man platoons each commanded by an officer; the platoon was divided into platoon H.Q. commanding three 10-man sections led by a non-commissioned officer. A section was known as a 'dropping' unit'.

The firepower of the battalion was an assortment of weapons – the hand grenade, sub-machinegun, pistol, rifle, Bren light machinegun, PIAT (Projector Infantry Anti-Tank) and two and three inch mortars (in the mortar platoon of H.Q. Coy.) with a small proportion of medium machineguns. They also used exotic explosive weapons like the Bangalore torpedo and Gammon bomb.

The original glider-borne formation was the 1st (Air-Landing) Brigade Group formed in late 1940, commanded by Brigadier Hopkinson and consisting of 1st Bn. Border Regt.; 2nd Bn. South Staffordshire Regt.; 2nd Bn. Oxfordshire and Buckinghamshire Light Infantry; 1st Bn. Royal Ulster Rifles, a reconnaissance company, an anti-tank battery and other brigade units. They wore their traditional badges on maroon berets. Each battalion consisted of four rifle companies each of four platoons, a total of about 976 officers and men. Not subject to initial selection procedures, air-landing troops underwent extensive infantry training and received an extra shilling a day special duty pay. Possessing a heavier scale of infantry equipment, their role was to reinforce parachute troops in battle and to fly in guns, stores and supplies.

In September 1943, the 2nd Bn. Oxfordshire and Bucks Light Infantry and the 1st Bn. Royal Ulster Rifles were transferred to 6th (Air-Landing) Brigade, organised and led by Brigadier the Hon. Kindersley; in time they were joined by 12th Bn. Devonshire Regiment. At the close of 1943, the 1st and 4th Parachute Brigades joined Brigadier Hicks' 1st (Air-Landing) Brigade to form 1st Airborne Division; their glider troops were 1st Bn. Border Regiment and 2nd Bn. South Staffs reinforced by 7th Bn.

Kings Own Scottish Borderers. Thus, six old established British infantry regiments flew in gliders during World War Two.

Although the R.A.F. assumed full responsibility for parachute training and the operational despatching of paratroops, the provision of glider pilots was an army responsibility, so the Glider Pilot Regiment was formed in December 1941 and consisted of two battalions by August 1942. Volunteer soldiers presented themselves for normal air crew selection and were trained on powered aircraft before being transferred to the Glider Pilot Regiment. Highly trained and superb in their bearing and turnout, glider pilots were an impressive body of men, selected after physical, psychological and moral fitness tests had established a failure rate of 25%. They had a 12/14 week pilot training in powered aircraft, followed by conversion courses from the Hotspur trainers onto the light operational Horsa gliders. They were awarded wings worn on the left breast when qualified pilots with six months specialised training and proficient in both day and night flying; the most experienced of them handled the heavy duty Hamilcars. The glider pilot was trained to fly light and heavy gliders, filled with soldiers or materials, behind tug-aircraft in all types of weather. When the tow-rope was released, in daylight or darkness, he had to pilot his engineless craft to a smooth landing on enemy-held terrain. On the ground glider pilots were trained to fight individually as infantry soldiers or collectively as a combat unit under the direction of the Operational Commander. Just before D-Day, the Glider Pilot Regiment adopted the R.A.F. system by renaming their Army battalions Nos. 1 and 2 Wings, consisting of nine squadrons, formed of flights of four officers and 40 other ranks, usually Staff Sergeants, or Sergeants. Each glider crew were teamed-up at the home airfield with its respective tug-crew. Commander Glider Pilots was Brigadier Chatterton.

Gliders were a one-shot weapon serving no useful purpose after the assault landing and their pilots, despite their extensive training, flew only one, two or three missions, being counterproductive at all other times when, to maintain their state of training, tug-aircraft had to be made available. These were factors which may have caused Field Marshal Montgomery to tell the

British Press that airborne forces were a waste of good men! And, when the casualties and mixed results are balanced against Britain's uphill fight in the Second World War, it could be claimed that the major British airborne programme was a waste if only because airborne divisions spent too much time out of action and diverted potential leaders to minimum roles. With hindsight, it can be claimed that Britain should have raised only two or three parachute/glider brigades for operations such as Bruneval and D-Day, and not have attempted to duplicate existing American formations in Europe. But in 1940/41 when the British airborne troops came into being, the British had no way of knowing that they might not have to bear the brunt of the re-invasion of Europe and, above all, Winston Churchill saw in parachute troops a return to the age of heroic combat. By the end of the war in 1945 the individual courage of the airborne forces had been proved often enough to nourish a legend which obscured their mis-use and frustration.

The largest single foreign element in the British Airborne forces were the Poles, the only exiles to be organised as a conventional formation – the Independent Polish Parachute Brigade, commanded by General Sosabowski. In 1941, the R.A.F. also had under instruction men from Czechoslovakia, Norway, France, Belgium, Holland, Germany, Italy and Spain; many of them were formed into special units trained to operate with Resistance Movements in their own countries.

Prior to the debacle of May/June 1940, the French Army had but two companies of parachutists who saw no action. Then, for five years French airborne units were raised and equipped by the British and Americans. In 1941 two Free French battalions became part of the Special Air Service Brigade, serving on special missions in France and with the British in North Africa; the 2nd jumped into France on June 5th and later fought in the Loire Campaign and at Bastogne. The 3rd were dropped in small groups to bolster French Resistance Forces in June and July, suffering heavy casualties.

About 100 Canadian soldiers received parachute training at Ringway, returning to Canada to recruit and train at their own parachute school in Manitoba. In July 1943 the 1st Canadian

Parachute Battalion returned to Britain and joined Hills 3rd Brigade in the 6th Airborne Division replacing the 7th Parachute Battalion, who were transferred to Poett's 5th Brigade to bring them up to full fighting strength.

American

In early 1940, the U.S. War Department were studying the feasibility of paratroops (air infantry) and the practical aspects of the air-transport of all types of ground troops, and their airborne programme proceeded throughout that year under the joint control of the Chief of Infantry and the Chief of the Air Corps. Undoubtedly stimulated by German successes as a U.S. Army Historical Division Report 1945, 'The Airborne Command and Centre', indicates:

'Probably the greatest single impetus to United States airborne development and expansion was provided by the German invasion of Crete in May 1941. Here, for the first time in history, airborne forces were employed en masse in a combined effort of major proportions. Prior to this operation little consideration had been given to the use of gliders or powered aircraft for the landing of ground troops, emphasis having been placed entirely on the development of parachute forces. Here was a conclusive demonstration of the ability of glider-borne troops to effect tactical landings bringing in with them heavy weapons and transportation essential to the success of sustained ground action in overcoming organised resistance. 'Airborne thinkers' seized upon this operation as an illustration of the unlimited capabilities of a balanced airborne force.'

In August 1940, a 28 man Test Platoon selected 200 volunteers, made the first parachute jumps at Fort Benning, from a B18 bomber. Wearing the standard Air Corps T3 ripcord parachute and with reserve parachutes on their chests, the men first jumped from 1,500 feet and then from 1,000. At this time the early training of paratroop instructors was aided by American fire-fighting personnel trained to parachute into remote wooded areas to combat forest fires.

On 16th September 1940, the U.S. War Department authorised

the formation of the 501 Parachute Battalion and, with them still under strength, in July 1941 the 502 Parachute Battalion was formed. These early U.S. airborne forces encountered the same problems as did the British, training and tactical development being hindered by shortage of transport and bomber aircraft, shortage of parachutes but no lack of volunteers. The Americans made calculated use of the tough-guy image to build up the necessary spirit and morale in their new units. Over the door of the Parachute School at Fort Benning was written: 'Through these portals pass the toughest paratroopers in the world.' The Provisional Parachute Group under Colonel Lee established training schedules, establishments and tactical doctrines, besides issuing the parachute wings and distinctive jump-boots that were to the American paratrooper what the diving eagle badge and the red beret were to German and British airborne troops.

So extensive were the shortages that, during the manoeuvres of September 1941, only a single company of paratroopers was dropped, their equipment landing later from the same aircraft which had to return to base and re-load. In the November manoeuvres, three drops were attempted – the first in front of the Press was a shambles, the second became a demonstration and the third, a surprise operation, was successful. At a subsequent inquest demands were made for better and more frequent rehearsals.

In the months before they entered World War Two, the Americans built-up their airborne capabilities and in July 1941 successful air-landings were made with Waco CG/4A gliders, and the first air-landing unit – the 550th Infantry Airborne Battalion – was raised. Despite limited available aircraft the American airborne forces endeavoured to train themselves in the methods demonstrated by the Germans in Norway, Holland and Crete, placing stress on parachute assaults to seize airfields for the landing of infantry and powered aircraft.

On 30 January 1942, the four existing parachute battalions were expanded into four 3-battalion regiments, and in March a glider battalion joined them, trained to use the Waco CG/4A glider. In the same month the American Parachute Command became the Airborne Command under General Lee who exchanged visits and information with General Browning in Great Britain, arising from this the Americans adopted the British policy of relying on paratroops and glider-troops rather than air-landing transport aircraft. The 2nd Bn. U.S. 503 Parachute Infantry Regiment came over and were attached to the British 1st Airborne Division.

The American airborne forces achieved full combat status in August 1942, when the recently formed 82nd Infantry Division was reorganised as the 82nd 'All American' Airborne Division and the 101st 'Screaming Eagles' Airborne Division, under the commands of General Ridgway and General Lee respectively. Each was to consist of two parachute and one glider regiment; four parachute regiments were already in existence and two glider regiments were raised. By the end of the war five American Airborne Divisions and several Independent Airborne Regiments had been formed; in the European theatre they were mis-dropped in North Africa, scattered by weather in Sicily but successfully operated in Normandy, Southern France, Eindhoven/Nijmegen and across the Northern Rhine. Although lacking the heavy weapons necessary for sustained combat, they were used for long periods as regular infantry in Italy, Normandy, Holland and at Bastogne.

Initially ahead of the Americans, by the time of the North African invasion in November 1942, the British had lost most of their lead in airborne warfare – yet British airborne forces saw wider operational experience than the Americans during World War Two.

On 20 June 1944, General Eisenhower approved the organisation of the 1st Allied Airborne Army, formed of British 1st Airborne Corps (1st and 6th Airborne Divisions); 1st Special Air Service Brigade and 1st Polish Independent Parachute Brigade Group; the U.S. 18th Airborne Corps (17th, 82nd, 101st Airborne Divisions and later the 13th) and some airborne Engineer Battalions; and the British 52nd (Lowland Division), changed from a mountain formation to an air-landing force; the 8th U.S. troop carrier command with 38th/46th Groups Royal Air Force.

This new Airborne Army was commanded by U.S. Lieut. General L. H. Brereton, with General Browning as British Corps Commander and Deputy Army Commander; U.S. 18th Airborne

Corps was given to General M. B. Ridgway, his command of 82nd U.S. Airborne Division going to the dynamic General Gavin.

In Europe in 1944, the 1st Allied Airborne Army, although capable of vitally contributing to a decisive offensive, lacked the ability to achieve singlehanded success as the Germans had done in Crete. This, coupled with a seeming inability to respond quickly to a fluid situation, caused General Patton to favour the allocation of parachute brigades or regiments to armoured or infantry corps, to be held just behind the battle areas ready for quick-reaction airborne drops in support of advancing divisions.

FACTORS CONTROLLING THE EMPLOYMENT OF AIRBORNE TROOPS

Throughout military history commanders have sought an open flank so that their army could pin down strong frontal forces whilst assaulting weaker flanks, but they were restricted to a two-dimensional warfare on their left and right flanks. The advent of aircraft and the landing of troops from the air opened up a third or rear flank. Endowed with a high and special degree of mobility airborne forces, able to fly over hostile ground defences, could envelop from the open or vertical flank. It became evident that parachute troops possessed certain favourable aspects:

1. They could be delivered by an existing troop carrier or bomber aircraft so obviating the need to build a costly single-purpose assault aircraft.
2. Troops and aircraft could be quickly ready for action.
3. Because they did not have to land to deliver their human cargo, aircraft could choose their height according to circumstances. The delivery of airborne troops into a battle area was only practicable if air superiority was held and, if much of the flight was over enemy territory, anti-aircraft fire was the danger. Surprise, and sometimes deception were vital factors in aiding airborne transport to reach their objectives.
4. An airborne force had the same mobility and radius of

action as the aircraft themselves, although the use of parachutes and gliders to deliver troops is a one-way system where soldiers arrive on the ground scattered and armed only with light equipment, their supplies and ammunition limited to what they could carry from one supply drop to the next. The German Fallschirmjager had a song that aptly expressed their style of warfare – 'Comrades, there is no going back!', which indicated that the tactical mobility of airborne troops is limited because it is difficult for them to re-deploy or withdraw from the battle without helicopters, vertical take-off aircraft or secure landing strips.

5. It is the number of troops that can be settled down on the first lift that determines the scale of operations and, although it was tactically valuable to be able to drop troops by parachute in darkness it was often impossible to air-land troops except in daylight. Shortage of troop-carrier aircraft prevented the employment of overwhelming numbers. The limited manpower involved had to be men of high quality and training, specialist fighting troops although not necessarily specialist airborne troops, which imposed a limitation upon their numbers and economical use.

All soldiers require physical stamina, it was particularly crucial for the parachutist who had to be agile in landing and because formations scatter in the drop so that the men had to be fit enough to assemble on the run and proceed quickly to the objective. Training placed special emphasis on a system that allowed no units to be ignorant of its objectives if its officers were put out of action – a critical factor to airborne forces whose losses during the early phases of an operation, particularly among officers, were often high. Expected to possess superior standards of initiative and to be capable of assuming responsibility of ranks vastly senior to his own, airborne soldiers were elite troops whose seeming wastage through enforced inactivity between operations was balanced by the enemy's knowledge of their existence which forced him to allocate large numbers of troops to guard targets potentially vulnerable to attack from the air. Early in World War Two it was obvious that reports of a sizeable landing of airborne troops in rear areas had a shattering effect upon defenders who were unable to make a balanced assessment of the situation; even today thirty-six years later it would still cause considerable dismay!

Airborne forces are best employed in self-sustaining airborne actions where success depends either upon early relief by ground forces (as at Arnhem) or by reinforcing themselves from the air. Holland, Crete and the British landings at Primasole in Sicily show the importance of timely reinforcements to airborne assaults by a portion of the total available airborne force held in reserve. Airborne forces isolated from help on the ground or without reserves to jump as reinforcements, can be overwhelmed; these highly mobile soldiers, lightly armed and equipped with limited ammunition and supplies, become increasingly vulnerable to attack by stronger forces as their strength is reduced by fighting, fatigue and isolation. The American Airborne Commander General Gavin has said 'The offensive ability of an airborne force can best be measured in terms of its ability to defend.' If airborne forces cannot be reached by ground troops within the planned period, the situation might be saved by parachuting reinforcements so that the operation assumes an independent character.

Generally speaking, an airborne penetration of enemy rear areas is only successful if it can inflict disabling injury before being neutralised and, although a quick and relatively cheap victory can often be achieved, against a first-class enemy on the alert, it can be a gamble. Airborne commanders cannot test enemy defences by ground patrols and, as shown at Crete and Arnhem, an airborne action is a jump into the unknown which is likely to encounter unforeseen difficulties and peculiarities. Allied airborne planners had a tendency to think in terms of bombing missions – that their success was assured if the force could reach the target area and be quickly set down. There appeared to be small heed of any possible opposition from German ground forces. This was exemplified at a pre-Arnhem planning briefing when General Sosabowski, the brave and experienced commander of the Polish Parachute Brigade, expressed his misgivings at what he considered a dangerously over-confident plan, by bursting out – 'But the GERMANS ... the GERMANS!'

Although an exercise in air-power to achieve a ground-force

purpose, airborne warfare is a combat form in its own right and not just ground warfare transposed; its success should not rely too greatly on conventional methods. The conditions of the D-Day operations meant that the airborne forces were not isolated for long, so did not have to rely on air support for heavy fire power. In some senses there arose a dangerous misconception, by airborne forces seeking to increase their close-support artillery so that they resembled standard infantry divisions. As the war progressed this caused Allied airborne divisions in Europe to be increasingly regarded and employed as close-support battlefield units.

Generally speaking, in the Second World War airborne theory varied widely from practice – for the Germans, operations in Norway, Holland, Belgium, Greece and Crete were extremely rewarding and well worth the cost. In Europe, the Allies usually working in relatively unfavourable circumstances, did not conform to any pattern in their employment of airborne troops. They were used to overrun airfields in North Africa, to take key points in Sicily, and as beachhead reserve supporters at Salerno; they were seaborne at Taranto, employed as a flank cover and diversion on D-Day and supplemented an assault in the South of France; created an abortive corridor in the 'Market Garden' operation and served as close-support troops to the Northern Rhine crossing. Mistakes and disasters in Sicily and at Arnhem, diverted resources and repeatedly cancelled operations might cause severe critics to claim that the Allied Airborne forces of World War Two were the counterpart of Haig's cavalry reserve in World War One.

In Burma and the Pacific, improvised airborne actions required a relatively small diversion of resources, and there were obvious advantages in the case of island landings. However, American commanders in the Pacific seemed to believe the high qualities of airborne troops as infantry out-weighed their special skill as parachutists and Marine parachute battalions, after dropping in combat, were used as conventional infantry and suffered heavy casualties. It is interesting to note that the post World War Two British Army shows more citations for the Parachute Regiment than for any other unit, yet in the vast majority of cases, these men went into action other than by parachute.

The World War One battle of Cambrai demonstrated that the combined power of tanks, infantry and artillery was far greater than the sum of their separate abilities. In the same way operations in Holland and Crete, the Markham raid in the Pacific and Wingate's operations in Burma confirm that air power when working in close co-operation with ground forces can convert ordinary infantrymen into supermen able to achieve objectives far beyond the reach of airborne forces, or ground-force units working separately.

DRESS, EQUIPMENT AND WEAPONS OF AIRBORNE TROOPS

GERMAN

DRESS

In the early years of German airborne development the older type parachute jump-suit, lacking pockets and with two zippers running diagonally up its front, was in use with the '1st Pattern' Luftwaffe smock, which had pockets and tailored step-in legs. The distinctive loose-fitting blue/grey waterproof gaberdine coverall, just over knee-length, could be fastened around the top of the legs to prevent ballooning that might foul harness straps. It had a full-length zipper, the sleeves buttoned at the wrist and two large pockets on thighs and both pockets on upper chest were also zip-fastened. The coverall was worn over the uniform and equipment, removed after landing and re-donned under the equipment. Camouflage-pattern jump-smocks were introduced in Spring 1940, two versions being worn at Crete – the green/brown 'splinter' and the 'splotch' types.

Regular Luftwaffe blue/grey trousers were long and loose like ski-pants, and gathered into ankle boots. On the right leg was a buttoned pocket holding a jump-knife.

German airborne troops wore a lightweight round paratroop helmet with a narrow brim and minimal neck-shield; of standard Luftwaffe blue/grey and usually covered by a tan camouflage

FRONT

DRESS AND EQUIPMENT OF THE

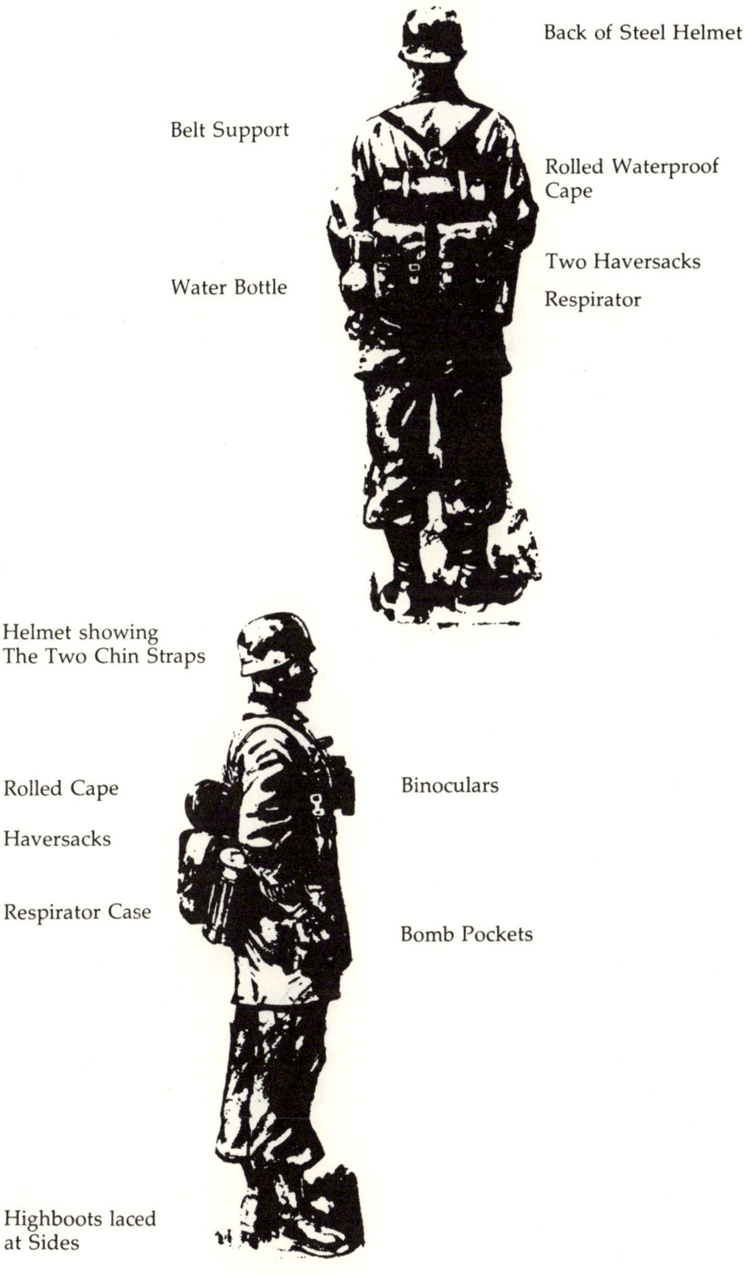

GERMAN FALLSCHIRMJAGER

cover with bands at sides and over the top to hold camouflage material. It was held on by a leather strap, forked below the ears.

Paratroopers wore padded leather gauntlets, knee and elbow protectors of thick horizontal rubber bars strapped over the coverall; all were discarded after landing. Paratroop boots, laced up the sides and with thick rubber soles, extended above the ankles; the trousers were gathered into them.

German parachute troops were distinguished by yellow Waffenfarbe on collar-patches and shoulder-straps; the 1st and 2nd Parachute Rifle Regiments and the Parachute Division wore a right-hand green cuff-band.

WEAPONS

9 mm Pistol 38 (Walther) Rugged and reliable, weighed 2·1 lb; magazine capacity 8 rounds; effective range 50 yards.

9 mm Pistol M1908 (Luger) A good weapon to fire and handle, the Luger was considered a prize trophy by Allied soldiers. Weight 1·93 lb; magazine capacity 8 rounds; effective range 55 yards.

Both used the 9 mm ammunition of the MP40 submachinegun, thus easing supply problems in the field.

7·92 Gewehr 33/40 A special shortened variant of this rifle was issued to parachute and glider troops. Weight 7·9 lbs; magazine capacity 5 rounds; maximum effective range about 600 yards. It had a powerful recoil and produced a great deal of muzzle-flash.

Mauser Model 98 Rifle Since its inception in 1880 this weapon in one form or another, was produced in larger numbers than any other military rifle. Calibre 7·92 mm; weight 8·75 lbs; magazine capacity 5 rounds; maximum effective range 600 yards plus. A good sturdy and accurate service rifle capable of using several types of grenade-launcher.

Maschinenpistole 38 One of the most famous of all submachineguns, the MP38 was mainly designed for airborne troops, with a folding butt and downwards pointing magazine. Known to the Allies as the 'Schmeisser', this and the MP40 were prized trophies of war. Calibre 9 mm; weight 9 lb; magazine capacity 32 rounds; cyclic rate of fire 500 rounds per minute; effective range 220 yards.

Maschinenpistole 40 Almost identical but far easier to mass-produce than the MP38 this was perhaps the best World War Two sub-machinegun – captured guns being used in preference to Allied equivalents. Calibre 9 mm; weight 8·87 lbs; magazine capacity 2 × 32 rounds; cyclic rate of fire 500 rounds per minute; maximum effective range 220 yards.

Light in weight, compact and easy to strip and service. Six extra clips of ammunition were carried, three to a pouch.

7·92 mm MG34 One of the two standard German machine guns of World War Two, the MG34 weighed about 75 lbs with tripod (less with bipod); feed either 75 round drum or 50 round belt; maximum effective range on bipod 600 yards, on tripod 2,000 yards. Rates of fire – 800/900 rounds per minute (cyclic), 300 rmp (practical on tripod) and 120 rpm (practical on bipod).

7·92 mm Fallschirmjagergewhr 42 The first true 'assault' rifle and one of the most remarkable designs to emerge from World War Two, revolutionising machine-gun design. Resembling a light machinegun with a permanent bipod, it also had a folding bayonet. Weight 9·94 lbs; magazine capacity 20 rounds; cyclic rate of fire 750/800 rounds per minute; maximum effective range 800 yards.

Both these machineguns were versatile weapons, able to fill the light role with a bipod mount, and medium/heavy roles with tripods; additional fittings allowed them to be converted to anti-aircraft weapons. Their ammunition was common with that for the Mauser Kar 98K rifle. Due to its high rate of fire, the M42 often required barrel changes, and a spare barrel in a container was carried with the gun.

The various components of these machineguns were gathered from weapons cannisters and assembled mounted on a tripod to serve in the role of a heavy machinegun; from other cannisters the crew obtained supplies of ammunition so that they could provide defensive fire for the air-dropping zone, if required.

Hand Grenades Stielgranate 24 weighed 1·31 lb and Eiergranate 39 weighing 0·75 lb. Each had a delay period of 4/5 seconds. Could be thrown 40 yards; blast-radius 15 yards.

Panzerfaust (infantry anti-tank projector) This hollow-charge projector had a maximum effective range of about 100 yards but was most efficient at about 35 yards when it could pierce 200 mm of armour. Approximate rate of fire – 10 rounds per minute. Crew of two.

81·4 mm sGrW34 Heavy Infantry Mortar Could be broken down into three parts for transportation. Weight 125 lbs; bomb weighed 8·71 lbs; effective range 420-2078 yards; blast radius 18 yards; rate of fire 23 rounds per minute.

German 81mm Grw34 heavy infantry mortar and crew.

81 mm 1eGrW34 Heavy Infantry Mortar As above but with effective ranges of 330-750 yards.

5 cm 1.gr.W 36 Light Infantry Mortar Trigger-fired. Weight 31 lbs. Maximum range 568 yards. Broke into two parts (barrel and baseplate) for transportation. All these mortars had crews of three men and minimum ranges of 50 yards.

German airborne forces also used the 8 cm Kurzer Granatenwerfer 42 (81 mm short mortar).

10·5 cm Gebirgshaubitze 40. L/30. Gun. Calibre 105 mm; weight of shell 32·65 lb; weight 3660 lb; maximum range 13,810 yards. Advanced design with split-trail carriage; a version with strengthened axle and special ammunition was designed for use with parachute troops.

7·5 cm Leichtgeschutz 40. L/10. Gun (Recoilless). Used at Crete, being dropped in four separate containers (later only two were used). It had tactical disadvantages and required special safety precautions when fired. Calibre 75 mm; weight of shell 12·85 lb; weight in action 321 lb; maximum range 8900 yards; could penetrate 50 mm of armour at 500 yards at 30° angle-of-impact.

10·5 cm Leichgeschutz 40. L/13 gun. (Recoilless). Designed primarily for the airborne role, it was dropped either in four containers or complete in a specially designed tubular frame. On the ground it was towed by a SdKfz 2 Kleines Kettenrad, the combination being capable of stowage in a Gotha Go 242 glider. Calibre 105 mm; weight of shell 32·6 lb; weight of gun 855·5 lbs; maximum range HE 8694 yards.

2·8 cm Panzerbuchse 41 (anti-tank gun). Although called 'a heavy anti-tank rifle', this was a gun in everything but its calibre, being the first weapon with a tapering-bore to reach the battlefield. The airborne version of this high-velocity weapon had a lightweight cradle and a pair of small aircraft-type wheels, with a single pole tubular steel trail; it lacked the shield of the standard model. Calibre 28 mm tapering to 20 mm; weight 260 lbs (standard model 505 lbs); weight of projectile 4·62 ozs; penetration at 100 m – 94 mm thickness at 0°; 69 mm at 30° angle of impact; at 500 m – 66 mm at 0°; 52 mm at 30°.

It is probable that German airborne forces also used the 3·7 cm Panzerabwehrkanone 36 (3·7 cm Pak 36); the 5 cm Panzerabwehrekanone 38 (5 cm Pak 38) and the 7·5 cm Pak 40 their weight and size would have required them to be brought in by glider. However, the Germans worked on the policy that any and every gun might be called upon to fight tanks, so that every gun was provided with some sort of anti-tank projectile.

BRITISH

DRESS AND EQUIPMENT

Largely copied from the Germans, the apparently complicated British parachute clothing was simple to wear and provided adequate warmth in the air and maximum mobility on the ground. The British airborne soldier wore standard khaki serge battledress with web waist-belt and gaiters, and ammunition boots. Over the battledress blouse (under which was a shirt and perhaps a pullover), he wore the Denison smock, a windproof garment camouflaged with an irregular pattern of brown and dark-green on a light-green base; it had large patch-pockets, zip-fastened right down its front, and included a 'tailpiece' drawn up under the crutch and snap-fastened at the front. Over the Denison smock, was worn a jump-jacket that covered the equipment to prevent anything catching in the parachute harness.

The paratrooper was proud to wear a maroon beret; on active service a protective rimless helmet with no neck-shield was worn, its strap embodied a chin-cup; it was covered by a dark-green string net garnished with foliage or coloured sacking-strips.

The pockets of the smock and trousers were packed with items of weaponry, grenades and spare magazines; sometimes a pistol was stowed inside his jacket. During the descent the Thompson or Sten submachinegun was fixed across the chest inside the upper strap of the parachute harness.

The basic British webbing equipment was the 1937 Pattern – standard issue small pack, supported by two shoulder-straps; a pair of ammunition pouches fixed to the waist-belt and secured to the body by a pair of narrow supporting straps. Also on the belt were waterbottle and carrier; bayonet and frog; entrenching-tool with haft and carrier, and a toggle-rope.

Heavier weapons, ammunition and supplies were carried in a 100 lb kitbag attached to the man's leg, hanging down below him on a long line when the parachute had opened; the British 'X' type parachute had gentle opening characteristics that allowed a man to jump thus encumbered.

WEAPONS

Rifle Short Magazine Lee Enfield Rifle (SMLE) ·303, magazine-fed, bolt-action shoulder rifle. 10 round magazine, 5 single shots per minute, normal, 15 single shots per minute, rapid. Muzzle velocity 2,440 feet per second; effective range 600 yards; weight 8·75 lbs.

Pistols Webley (Enfield) ·38 Revolver No.2 Mk.I. Weight 11 lb 11½ ozs. 6 rounds. Effective range – 20 yards (point-blank). Standard side-arm of the British Army from 1936—1957.

Smith and Wesson (U.S.) ·38 revolver. 6 rounds.

Webley (Br) ·45 revolver. 6 rounds.

Webley (Br) ·45 magazine-fed S.L. pistol. 7 rounds.

Colt (U.S.) Model A1 ·45 automatic pistol. 7 rounds. Weight 2.43 lbs. Effective at 50 yards.

Sub-Machine Guns

Thompson A1:·45 50 round drum or 20/30 round box magazines. 600/725 rounds fired per minute. Weight 11 lbs. The M1, introduced in 1941, dispensed with the drum-magazine.

The Tommy-gun was a clumsy weapon, reasonably accurate up to 50 yards.

The Sten 9 mm Machine-carbine replaced the Thompson. In five versions, with 32 round magazine, selective method of fire at a maximum rate of 500 rounds per minute. Weight 7 lbs. Effective range up to 100 yards.

No. 36M Mills Grenade. Weighed 1·7 lb and had a delay period of 4 to 7 seconds, made variable by the length of the fuse. Both grenades had approximate blast radius of 15 yards.

No. 69 Mark 1. Weighed 0·683 lbs and had an 'all-ways' impact fuse.

Flamethrowers, Infantry Fuel carried in a cylinder on man's back, and pumped out through a projector held in the hands. Range about 28 yards, tank capacity of ten 10 second discharges.

Machine-Guns

Bren ·303 Light Machine-gun. Air-cooled, gas-operated with either 30 round curved box magazines or 100 round-drum magazines; fired single or automatic – 60 rounds (normal); 120 rounds rapid, or maximum rate 450/500 rounds per minute. The Bren could be supported on the ground by adjustable bipod or tripod pivots, or fired from the hip. Effective range on bipod was 500 yards, maximum 2,000 yards. Weight 23 lbs.

Vickers Medium Machine-gun. Basic British infantry machine-gun. Water-cooled, belt-fed, belt contained 250 ·303 cartridges, fired at maximum rate 450/530 rounds per minute. With tripod, weighed 90 lbs; required crew of two to load and fire. Effective at 2,000 yards.

Anti-Tank Weapons

P.I.A.T. Projector Infantry Anti-Tank MkI, projected a 3 lb hollow-charge grenade, recoil forces re-cocking the firing-mechanism. Muzzle-velocity 240/450 feet per second, with effective range of 100/150 yards. Weight 32 lbs. 4″ penetration.

Grenades

Hand Grenade No.74 (St) 'Sticky Bomb' covered with adhesive to stick to tank-armour. Weighed 36 ozs, with 5 second fuse.

Hawkins Grenade No.75, using a crush-igniter, could be thrown as grenade or used as ground-mine. Employed in clusters against heavier tanks. Weighed 36 ozs.

Artillery

75 mm Pack Howitzer M1A1 on Carriage (Airborne) M8. American-made, this was the primary artillery support weapon of the airborne division. 6 rounds per minute; muzzle-velocity (with H.E. shell) 1,250 feet per second. Effective range 9,475 yards. All-up weight ½ ton.

40 mm BOFORS. Used as light-weight anti-aircraft weapon. Fired 2 pdr shell at 120 rounds per minute.

20 mm Hispano-Suiza (Oerlikon) was also employed for air-defence.

Mortars

OML 2in Mortar Mks VII and VIII. Quick into action, this muzzle-loaded support weapon fired 2¼ lb HE or 2 lb smoke-bombs with great accuracy over 500 yards at maximum range of 8 rounds per minute. Weight 19 lbs with base or 10½ lbs with spade.

OML 3 in Mortar Standard Medium mortar of battalion heavy weapon companies. Muzzle-loaded, fired 10 lb HE, smoke and illuminating bombs at 5 rounds per minute with an effective range of 1,600 yards. Weight bipod mounting 42 lbs, barrel 42 lbs, base-plate 37 lbs. Fragmentation radius 100 yards.

OML 4·2 in Mortar. Maximum effective range 4,000 yards. Weight 257 lbs.

No. 82 Grenade (Gammon Bomb). Plastic explosive charge in stockinet bag with detonator mechanism in screw-cap at neck. Used for demolitions and as anti-personnel grenade.

6 pdr Ordnance Quick-firing, Mk II anti-tank, on modified carriage Mk 3. The standard towed infantry-support anti-tank weapon of World War Two, its calibre was 57 mm (2·244 ins); shot weighed 6 lb 4 ozs; could fire 10 rounds per minute; muzzle-velocity 2,800 feet per second. Armour penetration – 81/83 mm at 500 yards. At Arnhem, 26 of these guns were landed and they were also used in the Rhine crossing.

British 6pdr anti-tank gun and ammunition limber drawn by jeeps – all capable of being carried in Horsa or Hamilcar gliders.

17 pdr Ordnance Quick-firing, anti-tank. Calibre 76·2 mm (3 ins); muzzle-velocity 2,900 feet per second; weight of shot 17 lbs (AP; APC; APBC; HE); Armour penetration – 120 mm at 500 yards. Able to penetrate most German armour at battle ranges, the 17 pdr was the only Allied anti-tank gun that could match Tiger and Panther tanks.

Light Tank Mk VII 'TETRARCH'. This armoured fighting vehicle was adapted for the glider-borne role, the Hamilcar glider being specially designed to carry it. An airborne reconnaissance regiment was formed as part of the 6th Airborne Division for the invasion of Europe and a squadron of Tetrarchs and Universal Carriers was taken in by gliders on 6th June 1944. Their contribution to the success of the operation was limited, partly because they were so few in number.

Specification: Crew: 2 (commander/gunner; driver)
Battleweight: 7·5 tons.
Armament: 1 × 2 pdr; 1 × 7·92 mm Besa machine-gun.

British 17pdr anti-tank gun and crew – could be carried in Hamilcar glider.

Jeep, carriers and carrier-drawn anti-tank guns as transported and carried by British Horsa or Hamilcar gliders.

Armour Thickness: Max. 14 mm
Min. 4 mm
Maximum speed: 40 mph (road)
28 mph (country)
Road radius: 140 miles
Ammunition stowage: 50 rounds 2 pdr.

Airborne forces used many ingenious adapted or specially designed pieces of equipment. The entire allocation was not necessarily air-lifted on all operations. One of the major items was the Wellbike two stroke motor-cycle, specially devised for paratroopers. Fitting into an equipment container, it was 4 ft 3 ins long × 15 ins broad and weighed 70 lbs. Handle-bars, steering column and saddle were on a collapsable principle. Maximum speed 30 mph; maximum range 90 miles on 6½ pints of petrol.

AMERICAN

DRESS AND EQUIPMENT

American parachutists wore a combined combat/jump jacket similar to the M 1943 standard infantry combat dress, but with four large patch pockets, and with a full-length frontal zipper. The trousers were of the same material, baggy and with a large thigh pocket on each leg; they tucked into the top of the jump-boots. On some occasions, U.S. airborne troops wore the ordinary issue fatigue uniform as a jump-suit.

The jump-boots were high, front-laced leather boots and, until the high lace-up combat boots replaced the infantryman's high shoes and canvas leggings, were the proudly worn distinctive mark of the paratrooper.

The American airborne helmet was similar to the M1 steel combat helmet, but with a moulded leather chin-cup that could be discarded on landing. Dark olive-drab netting embellished with scrim covered the helmet and also held a field-dressing in place.

The equipment was a woven waist-belt that supported a pistol-holster, a pistol ammunition pouch, a water-bottle in a canvas carrier, and an entrenching tool, a small pack and blanket-roll was carried on the back.

Alone of World War Two paratroopers, the Americans wore reserve parachutes; sometimes strapping a weapons bundle on the chest below it – the rifle was carried strapped to the side during the drop.

WEAPONS

Pistol, automatic, calibre ·45, M1911A1. This Colt pistol had a magazine capacity of 7 rounds and weighed 2·44 lb. Maximum effective range 70 yards.

Rifle, calibre ·30, M1. The Garand M1 rifle has the distinction of being the first self-loading rifle to be taken into service as a standard weapon by any service-arm. It was 43·6 ins long, weighed 9·5 lbs; magazine capacity – 8 rounds. It could be used with a bayonet. Maximum effective range 2,000 yards.

Reising model 50 sub-machinegun. Not greatly used, this weapon had a calibre of ·45 in; length 35·75 ins; weight 6.75 lbs; magazine capacity 12 or 20 rounds and cyclic rate of fire of 550 rounds per minute. Maximum effective range – 650 yards.

Sub-machinegun, calibre ·45 in M2. Accurate and easy to use but difficult to produce, only a few were issued to airborne forces. 32·1 ins long; weighed 9·25 lbs; magazine capacity 20 or 30 rounds; cyclic rate of fire 500 rounds per minute. Maximum effective range – 650 yards.

Sub-machinegun, calibre ·45, M3. Closely following the Sten concept and possessing the same reliability and efficiency, this was the principle automatic weapon of the U.S. airborne forces. It was 29·8 ins long; weighed 8·15 lbs; magazine capacity 30 rounds; cyclic rate of fire – 350/450 rounds per minute. Maximum effective range – 650 yards.

Sub-machinegun, calibre ·45, M1 and M1A1. Often preferred to the M3, this was a modified Thompson, 32 in long; weighing 10·45 lbs; magazine capacity 20 or 30 rounds; cyclic rate of fire 700 rounds per minute. Maximum effective range – 650 yards.

Grenades. In World War Two the Americans used two types of grenade – the Mark 2A1 Fragmentation grenade weighing 1·31 lb; with a delay period of 4—4·8 seconds, and the Mark 3A Fragmentation grenade, weighing 0·84 lb and with a delay period of 4—5 seconds. Blast radius approximately 15 yards.

Browning Automatic Rifle. M1918A2 (BAR). Can be regarded either as a heavy automatic rifle or a light machinegun. Calibre ·30 in; length 40 ins approx.; weight 19 lbs; magazine capacity 20 rounds; rate of fire 550 rounds per minute; maximum effective range 2,000 yards.

7·62 mm M60 GPMG. Similar in type and conception to the German MG34 and 42, this General Purpose Machine Gun was 43 ins long; weighed 23 lbs; link-belt feed; cyclic rate of fire 550 rounds per minute; maximum effective range – on bipod 880 yards, on tripod 2,000 yards.

·30 Browning M1019A4 and A6 machine guns. Water and air-cooled respectively, this was a reliable weapon. Weight 32 lbs; 250 round belt feed; maximum effective range 1050 yards; cyclic rate of fire 400/550 rounds per minute.

Browning ·5 M2 Machinegun. M2HB. More of these guns have been made than any other American machinegun. Weight 85 lbs; 110 round linked-belt feed; maximum effective range 1100 yards; cyclic rate of fire 450—575 rounds per minute. Because of its weight, the ·5 m/g was more likely to be used from a vehicle pintle than a ground mounting.

2·36 in M9A1 Rocket Launcher (Bazooka). This two-man anti-tank weapon fired a hollow-charge projectile at relatively short ranges – its maximum effective range being about 120 yards. Ten shots per minute. Armour-piercing capability of 120 mm.

Mortars

60 mm M2 fired a bomb weighing 3 lbs; effective range of about 330 yards to 1250 yards (minimum 100 yards). Rate of fire – 18

American light paratroop mortar and crew.

rounds per minute. Blast radius 12 yards.

81 mm M1 fired bomb weighing a little more than 10 lbs; effective range of 440 yards to 1750 yards (minimum 100 yards). Rate of fire – 18 rounds per minute. Blast radius 17 yards.

75 mm Pack Howitzer M1A1 on carriage M8. Used carriage (airborne) M8 with rubber tyres that could be easily adapted for dropping by parachute – with gun it weighed 1,340 lbs; the equipment could be broken down into 9 loads for parachute delivery. Used by British Airborne Artillery units, who knew it as the 75 mm Pack Howitzer Mark 1. Shell weighed 13·76 lbs; maximum range 9,475 yards.

105 mm Howitzer M3 – Carriage M3A1. Serving only with American units, this shortened version of the 105 mm M2 Field Howitzer was designed specially for airborne and jungle warfare. Weight 2,495 lbs; shell weighed 33 lbs; maximum range 7,250 yards.

57 mm M1 Anti-tank gun. Effective range 1,100 yards (armour piercing) and 1,300 yards (HE); could penetrate 68 mm of armour at 500 yards; capable of firing 15 rounds per minute.

3 in (76·2 mm) M5 Anti-tank gun. Effective range 1,100 yards (armour-piercing) and 1,300 yards (HE); could penetrate 90 mm of armour at 500 yards; capable of firing 18 rounds per minute.

PARACHUTES

Used in World War One by balloonists hastily evacuating blazing balloons, military parachutes were developed in the years between the wars, first by the Italians, then the Russians and Germans. In Britain, the Royal Air Force provided the first airborne troops with standard rip-cord parachutes until in early 1940 it became apparent that the Irvin trainer parachute could not be satisfactorily adapted for automatic pulling of the rip-cord, one pupil being killed and three more failures occurring when parachute-fitted dummies were test dropped. Although the parachute worked perfectly well as an emergency 'chute when manually operated, the automatic method of opening could cause the canopy to tangle in the rigging lines. With impressive speed Raymond Quilter of the Gregory-Quilter Parachute Company produced a parachute where the rigging-lines were withdrawn from the pack with the canopy still folded in it, as the jumper fell to the limit of the lines the canopy was pulled clear of its pack and opened. A bar ran along the roof of the aircraft's fuselage with strap attachments to which the parachutist hooked a static-line which, once fully extended by the man's weight after he had left the aircraft, broke open the bag to release the rigging lines and the canopy. The man himself was attached to these rigging lines by harness lift-webs and after he had gone, the bag was left dangling on the end of the static-line and was pulled in by the despatcher.

Known as British Type 'X' Equipment, popular for its reliability and absence of opening shock, this parachute was the best of the wartime models, *allowing jumps to be made from tactically desirable heights of about 400 feet. Improving dropping accuracy, this also meant that the men were only in the air for a short period of about 40 seconds, so minimising the chances of casualties from enemy fire. The parachute measured 28 feet across when laid out with a centre circular vent 22 inches in diameter; earlier types were made of silk but later 'Ramex' cotton parachutes were introduced and nylon canopies were manufactured towards the end of the war. The parachute had 28 silk or nylon rigging lines each 25 feet long with a minimum breaking strength of 400 lbs; they were brought down in four groups of seven lines. The canopy was packed in an inner bag and an outer pack.

The straps of the parachute pack and harness were tightly but comfortably adjusted to bring the top of the pack squarely in line with the shoulder blades. Held in position by shoulder, back, chest and leg straps, the man sat in a 'seat' formed by the main suspension straps; metal strap attachments clipped into a metal 'safety' box attached to the chest straps. The man locked his harness by giving the circular metal disc controlling the box device a quarter-turn in a clockwise direction; this disc was given a sharp knock by the stick commander during his pre-emplaning inspection as he checked that it was closed and that straps were correctly adjusted.

From 1941 onwards, U.S.paratroopers used the T5 or T7 'chute, mostly the latter which was strong and reliable but with a large canopy and a slow rate of descent resulting in a straight and steady flight. Both the T5 and the T7 used a three-point harness fixing, which gave a severe jerk on opening besides being slow in release on landing.

Russian and American paratroopers wore a reserve parachute so that training jumps had to be from above 800 feet to allow for the use of the second parachute in an emergency; this must inevitably result in lack of confidence if an operational jump has

*The best military parachute so far devised, every paratrooper today uses a parachute derived from the 'X'type, or a straight copy of it.

to be made below the height at which a reserve 'chute can be used.

The normal jump height for German paratroopers was about 400 feet, allowing for a quick descent that reduced exposure to enemy ground fire. Dropping at such a low altitude required a 'chute capable of opening automatically through a static-line. Model Rz1 was the first such parachute, but it was not particularly reliable and the later RZ20 was most used during World War Two. The R2 parachutes opened canopy-first – the taut static-line jerking open the bag so that the canopy came out into the aircraft's slip-stream, filling with air as the man fell away beneath it to pull out the rigging-lines from the pack. They became taut with the canopy half-filled, bringing the jumper up short with a bruising jerk that was very quick but highly uncomfortable. The harness arrangement converged all the lines onto a single ring attached by two ropes to the back of the harness just above the man's waist causing him to hang slightly face-downwards with no control over his parachute; he turned himself in mid-air by making swimming movements with his arms. Landing in this position and in a fast forward direction was difficult and required a fit and agile man to do a forward somerault on hitting the ground. Although padded-up, injuries were common.

The lowest possible operational dropping heights were maintained to prevent the widespread dispersal of troops and their equipment on the ground – even under favourable conditions the stick-length* between the first and last men on the ground of a quick Whitley ten-man stick could be as much as 500 to 600 yards. The aircraft was liable to lurch and sway as it descended to the prescribed dropping-height and reduced speed to 100/120 m.p.h. for the run-in and actual dropping. The general exit routine was initiated by the orders 'Hook-up!' and 'Prepare for Action!'; on the red light signal came the command 'Action Stations' as, under the watchful eye of the despatcher, No. 1 moved into the exit hole or the door, closely followed by the remainder of the stick. On the green light signal of 'Go', depar-

*The term 'stick' defined the number of men to jump consecutively as a team from aircraft, and 'stick-length' was the distance between the first and last man to land; a stick of ten men was equivalent in size to an infantry section.

tures were made in an upright position with each man being pushed feet first from the door of the Dakota. This was in marked contrast to the German head-and-arms first into the slip-stream method of leaving the Ju52.

Paratroopers exited from the American Douglas Dakota C47 from the door but exits from British converted bombers were almost always made 'through the hole'. Holes came in different sizes and shapes according to the aircraft, the Albemarle hole was coffin-shaped and larger than the Whitley hole. Accidents sometimes happened despite the intensive training that taught every paratrooper that the safest way of jumping, both through the hole in the floor or the door, was to hold the body stiffly at attention. If the body entered the slipstream at any other angle or somersaulted forwards, the rigging-lines inevitably twisted together so that the front and back sets of rigging-ropes became interwoven to prevent the canopy opening properly. When this occurred, the rate of fall increased and no flight control was possible until the twists in the rigging-lines were eliminated and the canopy allowed to open fully. A dangerously fast descent followed a 'thrown line' or blown periphery, when part of the canopy periphery was first blown inwards and then outwards through the rigging-lines to produce a secondary inverted canopy. Or, a portion of the canopy could blow between two rigging-lines and roll up at the skirt. When a parachute canopy paid out but failed to develop until the last moment it was known as a 'streamer' and the most feared of all freak conditions was the 'Roman candle', the 'streamer' that failed to open at all.

It is recorded that over 400,000 drops were mounted from Ringway training airfield during the war and only one fatal accident was traced to faulty packing.

When a parachutist left an aircraft, the effect of the plane's slipstream was modified by the drag of his static-line, so that he was whisked past the tail of the aircraft as though supported. In fact his canopy was open before much vertical height was lost and before taking up a good parachuting position he had enough time to go through his training drill sequence: looking up to see if the canopy had opened normally; to turn about in the harness and make an all-round observation, taking evasive action if floating on a collision course with another parachutist. Once the parachute canopy was fully developed, the parachutist could look up and see that he was suspended from two harness buckles on his shoulders by four 3 foot webbing straps (lift webs) each attached by a metal ring to seven rigging-lines rising 25 feet upwards to the periphery and into the fabric of the parachute canopy. Two of the four lift-webs inclined forward and two backwards from the shoulder buckles and he could reach up and grasp them to achieve an appreciable degree of flight control by pulling down on the appropriate straps.

In the early days of parachuting it was discovered that by pulling down on either one or two of the lift-webs the shape of the canopy could be altered like the sail of a ship and resistance to wind currents increased, so that a fast forward drift was corrected by manipulating the back lift-webs and vice versa. In time, men learned the art of arresting the speed, drift and oscillation of a parachute.

The kitbags and weapon-valises dangling beneath the parachutist lessened the shock of landing, even so the average force of impact could be compared to the shock of leaping from a 12 foot wall. Parachutists were taught to swiftly disengage on the ground from the harness straps, to avoid being dragged along if the wind inflated the canopy; taught to safeguard the face when landing in trees and, if landing in the sea or a river, to completely release themselves from the parachute harness before entering the water.

Paratroopers were burdened with Bren or rifle valise, and an airborne kitbag weighing about 60 lbs attached by a rope to the lower harness-strap, and secured by an ankle-strap to the right leg. Stowed in an external pocket of the kitbag, the 20 foot length of rope paid-out as the bag was released in flight by jerking out a pin on a cord attachment from the ankle strap; a spring device absorbed the shock of release of the heavy bag. At Arnhem, many of them broke away from their bearers and smashed to the ground destroying such precious items as radio-sets. The Bren and rifle valises were clutched tight to the body on exit, and were attached to the harness in the same way and also released from the ankle-strap.

GLIDERS

The Russians began the development of the military assault glider but the Germans were the first to use them under actual combat conditions in the initial airborne action ever undertaken against really determined opposition, the glider-borne attack on the Belgian fort of Eben Emael, which indicated the complete faith of General Student and German senior airborne commanders in the untried glider technique. The parachutists of Sturm Abteilung Koch trained six months for their attack in DFS 230 Assault Gliders. The first assault glider in military history, it carried a pilot, a co-pilot and nine fully equipped men, who could exit rapidly through doors at each end. On operations, a Ju52 towed one glider, three Ju52's with their gliders flying in formation.

Specification

Weight loaded – 4,600 lbs.
Towing speed – 131 m.p.h.; Maximum speed – 181 m.p.h.

The DFS230 was developed as a means of delivering an assault from the air without separating the troops from their weapons; it could carry heavier armament directly into battle. This enabled the German paratroopers to be safely landed fully equipped with MP40 sub-machine guns and MG34s ready for action, obviating the need to find a weapons cannister. This glider possessed a defensive armament of a single 7·92 mm machine gun 15 mounted on the upper deck of the forward fuselage, with a flexible mount allowing it to give supporting fire to the occupants of the glider; it could be removed and taken into the assault. Eventually, two MG44s were fixed to the nose of the glider for increased fire power. Later, the DFS230 glider had a parachute-packing stored in the tail section of the aircraft, allowing it to dive at a steeper angle; then forward-firing braking-rockets were fitted to the nose – both devices permitting a much shorter landing space. The gliders used on the Gran Sasso Raid were fitted with these items.

The Germans also used the *GOTHA GO242*, a twin-boom, rear-loading machine, smaller than a Horsa but capable of carrying a 75 mm anti-tank gun and crew. This glider had diving-brakes of retarding 'ribbon' parachutes packed in its tail. Released at about 13,000 feet and 20 miles before the objective, the glider flew until the pilot estimated that the landing point was nearly directly below, then tipped the machine through 80 degrees and dived almost straight down towards the earth. Without special equipment, the glider would have disintegrated after exceeding its safe maximum speed, but the parachute brake held the speed at 105 knots. At about 800 feet the glider pulling out of its dive, reduced speed and circled the landing site before touching down. Flak was avoided by spinning, or diving by steps to change the angle of dive, a daring technique requiring a high

D.F.S. 230

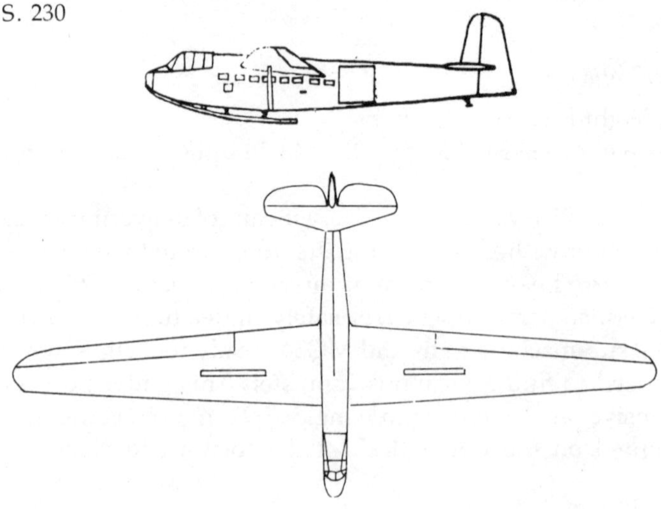

degree of skill and training. The glider's wheels were jettisoned after take-off, the landing run being shortened by wrapping barbed wire around the landing skids, or by firing a retarding rocket fitted in the nose of the glider.

Between 1941 and the end of the war Britain built over 7,000 gliders and trained 3,500 Army glider pilots, yet when the first 400 prototype Hotspurs were ordered by the Ministry for Aircraft Production, no such military machine as a glider existed in England, so one had to be designed to carry weapons, equipment and troop reinforcements. British gliders were:

GAL HOTSPUR

The early British-made glider, largely used for training purposes.

Specification

Weight loaded – 3,635 lbs. Six soldiers.
Maximum towing speed – 150 mph; Maximum diving speed – 170 mph.
Stalling speed – 54 mph.

AIRSPEED HORSA

Flying under both British and American colours on the Normandy invasion, the Horsa normally carried 15 fully armed troops, or alternatively a wide variety of military equipment stowed in the main compartment; the Mk II had a hinged nose to allow direct loading and unloading of light weapons and vehicles. When America entered the war an early model jeep was shipped to England and loaned to Headquarters 1st Airborne Division who quickly appreciated its potential for airborne operations. When tested with a mock-up of the Horsa glider at Ringway, the jeep fitted, leaving room for medium anti-tank gun or light field howitzer.

Specification

Weight loaded – 15,500 lbs.
Maximum towing speed – 160 mph; Maximum diving speed – 180 mph;
Stalling speed – 60mph.

Waco CG–4A

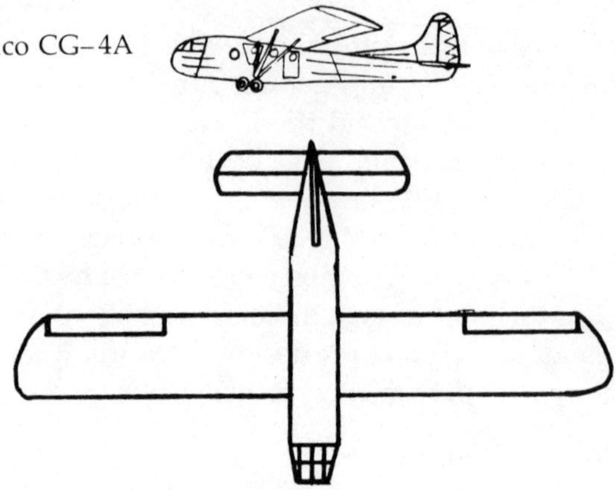

Airspeed Horsa

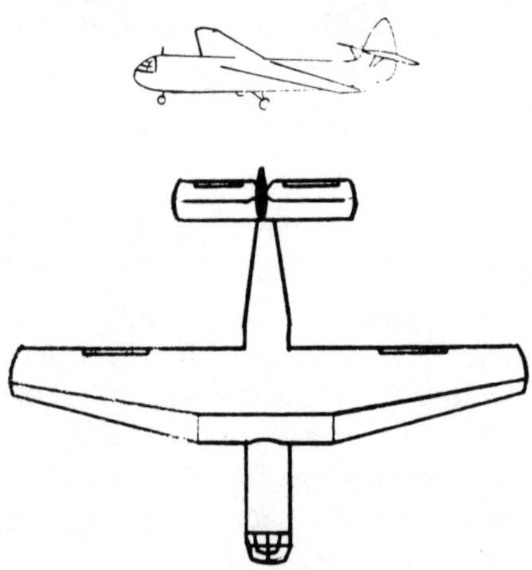

GAL HAMILCAR

This heavy-duty glider was originally designed to carry the Tetrarch tank or two Universal carriers, subsequently it was adapted to carry a great variety of military loads. Towed by Halifax, Lancaster or Stirling four-engined bombers, it was a high-wing cantilever monoplane made of wood and metal. The nose of the fuselage was hinged to starboard for loading.

General Aircraft Hamilcar

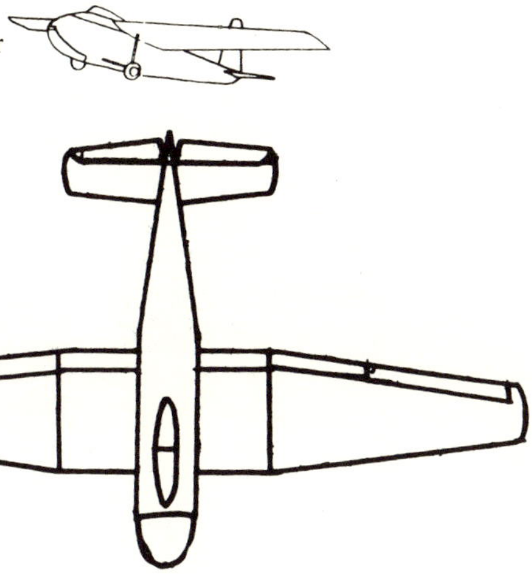

Specification

Weight loaded – 36,000 lbs.

Maximum towing speed – 150 mph; Maximum diving speed – 187 mph;

Stalling speed – 65 mph.

The Whitley, unable to tow a loaded Horsa, was limited to eight seater Hotspurs; the four-engine Halifax bombers originally allocated to tow the tank-carrying Hamilcar also had to haul the Horsa.

WACO CG-4A (British name Hadrian)

This American designed and made glider carried British airborne troops during the Sicily operations. It could carry freight or 15 fully armed troops, two being pilot and co-pilot. The nose was hinged for direct loading of equipment.

Specification

Weight loaded – 7,500 lbs.

Maximum towing speed – 125 mph; Minimum gliding speed – 38 mph.

The C47 tug-plane could carry 19 paratroopers against the Waco's 15, so that para-drops offered a numerical advantage which was reversed for the British whose Horsa gliders could carry 29 soldiers, and the large Hamilcars even more, although mainly used for such heavy equipment as anti-tank guns, vehicles and light tanks. Gliders could transport these items and other heavy equipment, although at the expense of having fewer soldiers for fighting purposes. Carrying anti-tank guns and light vehicles had to be balanced against the disadvantage of paratroopers being lightly armed and unable to cope with enemy armour; on the other hand it made them highly mobile and completely independent of roads and tracks.

One marked aspect of airborne operations is that there is rarely sufficient aircraft for every fighting man and his equipment, plus vehicles and anti-tank guns to be delivered in one lift. On the few occasions when this is possible and objectives won then on future operations the successful commander will be tempted to clamour for more aircraft!

On take-off the glider, on the ground, was pulled forward by the tug until the tow-rope was taut and the two aircraft climbed steadily into the air. During flight the glider's normal position was just above or just below the slip-stream of the tug.

When a glider pilot spotted his landing zone he readied himself to hear confirmation of location from the tug's navigator – 'O.K. number three . . . when you are ready . . .' The glider pilot acknowledged and the tug-plane's navigator wished him luck; then the glider pilot cast off (the tow-rope release mechanism was operated from the glider cockpit) and the tug disappeared with the tow-rope flapping in its wake. Eventually it was dropped on the enemy. The glider's airspeed fell off as the landing zone loomed nearer but the flimsy craft still came in at 60/100 mph; the co-pilot pushed his lever when asked for half-flaps and the glider bucked as the great flaps descended from underneath each wing and braked against their speed. With tree-tops leaping towards the floorboards and past the wings, the glider continued its descent, with the ground rushing up nearer until the pilot pulled back on the wheel, levelling the glider which hit once, bounced about three feet and came down on the ground. The brakes were slammed on and the glider careered across the grass until the wheels sank into soft soil and they ground to a halt, perhaps 50 yards short of solid-looking woodland. Running out of wind, some gliders came down in trees which ripped through their floor, nosing them over so that the glider flopped down to a loud background of snapping wood. Other gliders had their wings slashed off by the trees they careered through, and some came in at such a speed that they could not stop and were totally destroyed by tree trunks and thick branches as they ploughed into them. Heavy equipment – howitzers and jeeps broke from their chain-moorings to crush the gun-crew, pilot and co-pilot.

The huge eight-ton Hamilcars came to grief on landing in soft fields, their nose digging up earth in front of them until the weight and ground speed drove the nose deeper and the huge tail rose up in the air and the glider flipped over on its back. The Hamilcar had a hump where the pilot sat so that a flop-over onto its back usually killed him; the Horsa was flat on top.

The moment the glider came to a halt, the occupants poured out to unbolt the tails of the craft and swing them back to unload artillery pieces, equipment, stores, jeeps and trailers. The glider tail was held on by eight pins with a protective wire and in practice it was usual to get the tail off and the jeep and trailer out in about two minutes, but in action it did not seem to work so that, even after the wire was cut and the pins out, the tail would still refuse to budge until it was chopped off or a hawser attached to a jeep yanked the tail off.

In the Market Garden Operation the skilled pilots of the Royal Air Force and the Glider Pilot Regiment brought their craft over the dropping zones with clocklike precision and then, as the gliders cast-off, the tug-planes made climbing turns to allow space for those coming up behind. It was difficult to speed up the landing of gliders, which took three or four times longer to cast off than it took to drop an equivalent parachute unit.

The tow-rope that connected tugs and gliders required extensive longitudinal spacing; close formation flying was impossible because of the glider's inability to maintain an exactly true course. Gliders had to lose height and, whenever possible, approach the LZ into the wind which made for a slow approach when they were vulnerable to anti-aircraft fire. Small glider-borne units could achieve maximum shock-effect when used ahead of the main force but when used in large numbers their supplies potential was less than that of paratroops.

GLIDER-TUGS AND CARRIER-AIRCRAFT

The main essentials for airborne-warfare are eager battleworthy soldiers led by a commander with drive and imagination, *plus* a fleet of transport aircraft. Because of the last requirement, airborne warfare must be closely associated with the development of military air transport, and in 1940 the R.A.F. possessed no transport planes suitable for use as troop-carriers or for air-supply. Converted bombers such as the Whitley could only carry ten men or a limited amount of battle equipment. Even if the largest bomber were converted to carry 20 parachutists, it would need six aircraft to lift a rifle-company, and an airborne division with an establishment similar to a wartime infantry division would require nearly 900 aircraft, without any provision for its equipment.

Air Chief-Marshal Sir Arthur (Bomber) Harris came out strongly against any plan to divert his bombers to troop-carriers in support of large-scale airborne actions, considering that other than against practically unarmed and unprotected troops, neither parachute or glider operations were practical operations of war. In any event he believed the weather in Europe to be unsuitable for such operations; that enormous casualties would be incurred, and that such actions were unlikely to contribute materially to victory, as only one brigade could be carried at a time. In deference to his belief that to set aside bombers for transport duties would cripple the Bomber Command offensive, the Air Ministry issued orders that a number of aircraft and personnel were to be posted to R.A.F. units supporting airborne forces. But it was decreed that all bombers coming off production were to be equipped for glider towing, and Stirlings, Halifax's and Albemarles were to be modified for parachute dropping.

In the early days, lack of aircraft made training very slow because of the insignificant number of Whitley bombers allocated to the Training Establishment. This aircraft was thoroughly disliked, having neither the internal capacity or the right dropping facilities for training let alone operational needs. Nevertheless, considerable progress was made in the techniques of dropping and in the actual parachute equipment. Then, in early 1942 No. 296 (Glider Exercise) Squadron and No. 297 (Parachute Exercise) Squadron were formed; Whitley bombers being the principal aircraft employed by both squadrons. Number 38 Wing R.A.F. were assigned to lift 1st Airborne Division on training exercises and to instruct Bomber Command Pilots on dropping parachute troops and towing gliders. While Bomber Command was responsible for supplying aircrew and aircraft as required for operational duties, eventually Number 38 Wing flew its own missions and by Spring 1944 re-named 38 Group, had six Stirling, two Albemarle and two Halifax front line squadrons.

During World War Two, the Allies used the following aircraft as glider-tugs and carriers.

ARMSTRONG WHITWORTH WHITLEY

Originally a bomber-reconnaissance aircraft, the Whitley carried out the first paratroop operation over southern Italy in February 1941. The Mk V was converted into a paratroop aircraft with ten men dropping through a hole in the floor of the fuselage. It also served as a glider tug.

Specification

Maximum speed 230 mph; range 2,400 miles.

ARMSTRONG WHITWORTH ALBEMARLE

A twin-engined bomber-reconnaissance aircraft, the Albemarle was widely used as a glider-tug; it saw service as a paratroop aircraft with ten men dropping through a large hole in the rear fuselage floor. The Mk I, II, V, VI and Mk S.T.I. were all equipped as glider tugs; the Mk II, Mk V and Mk VI were alternatively paratroop-carriers; the Mk S.T.I. and the Mk VI were special transports.

Specification

Maximum speed was over 250 mph; range 1,350 miles.

HANDLEY PAGE HALIFAX

A four-engined heavy bomber, the Halifax was in continuous service until the end of the war, besides being allocated to glider squadrons and occasionally flying as a paratroop aircraft from some overseas stations.

Specification

Maximum speed 270 mph; range 3,000 miles.

SHORT STIRLING

Britain's first four-engined heavy bomber. By 1944, after five versions, the aircraft was obsolescent and was adapted as a transport, paratroop and tug aircraft, but was best used as a parachute supply aircraft. The Mk IV dropped 24 paratroopers from a large opening in the underside of the rear fuselage, bomb-cells were retained and used for the carriage and dropping of airborne supplies. Alternatively, 34 air-landing troops could be carried with arms and equipment; the Mk IV was also fitted for glider-towing. The Mk V was equipped for a variety of duties including troop transport for 40 fully-armed troops, paratroop transport for 20 men and containers, heavy freighter to carry one jeep, trailer and six pdr gun, or two jeeps with crew of eight men.

Specification

Maximum speed 280 mph; range 3,000 miles.

The Americans possessed the DC (Douglas Commercial) III which gave them an advantage because, as the most advanced civil transport of its day, it became the most successful military troop carrier of World War II, out-classing even the Ju52. It became apparent that the C47 was highly suitable as a paratroop-dropping aircraft, so increased supplies were requested under Lend-Lease but despite the high output of American aircraft factories, there were never enough C47s to satisfy both American and British needs. At times this led to bad feeling when British airborne forces had at their disposal more C47s than did the Americans who made them!

THE DOUGLAS C-47 SKYTRAIN, known to the British as the DAKOTA, was a twin-engine low-wing monoplane that combined advanced design with essential simplicity and performance with robustness. Capable of operating off improvised strips, it was equipped with folding benches for 19 fully-armed paratroops who exited from the port side aft door, or 27 infantrymen or 10,000 lbs of cargo. From mid-1942 the Dakota was the workhorse of British and American parachute forces, and also served as a glider-tug.

Specification

Maximum speed 229 mph; range 1,500 miles.

Before World War II the German air-line Lufthansa was equipped with the JUNKERS JU 52/3 mg7e TRANSPORT AIRCRAFT, a machine suitable for conversion to bombing or military transportation and already in mass-production at the outbreak of war. This aircraft, the standard transport of the Wehrmacht, made German airborne operations possible because of its availability in large numbers. Capable of operating from rough strips on roads or fields, it could carry 18 fully-armed troops or 10,000 lbs of equipment and was able to tow gliders at low speed.

Specification

Maximum speed 164 mph; cruising speed 131 mph; range 3,000 miles.

PATHFINDERS

Protected by darkness, airborne operations can be undertaken that would be unacceptably hazardous in daylight but require accurate navigation and a means of identifying the exact delivery area. At first, dropping-zone marking teams were dropped by the best qualified air crews to set out beacons for the mainstream that followed but in time came an obvious need for an airborne unit specially trained and equipped to 'act as the tip of the spear in airborne landings'. The dropping-zone marking parties of the Army Pathfinder Units experienced considerable difficulty even when correctly dropped because the ground beacons and the radio receiver sets in the aircraft only brought the troop-carriers into the area of the dropping-zone and the final choice of the release point had to be made visually. After Normandy, there was a certain crisis of confidence within the U.S. Airborne forces as to the ability of their carrier-aircraft to drop them accurately into battle.

So the Pathfinders of the Independent Parachute Company came into existence, having to be dropped and in position shortly before the arrival of the first lift so that they could provide ground aids or beacons to direct aircraft onto the dropping or landing-zones. Rebecca-Eureka, a simple air-to-ground radio device was employed by airborne forces – Rebecca being used to obtain signals from Eureka as a navigational aid as well as identifying airborne target zones. In daylight, dropping and landing-zone recognition was comparatively simple, but night operations presented location hazards. The usual method of dropping parachutists in darkness was for the carrier-craft to fly in a line marked by two lights placed several hundred yards apart and, as the ground lights appeared at 90 degrees to port and starboard on the line of approach, providing the aircraft was flying onto the dropping-zone on the correct bearing, the green light in the aircraft was switched on and the troops exited. Gliders used a flare path laid out in the shape of a T; pathfinder troops placing three electric lights 75 yards apart at the head of the landing strip with five more forming the stems some 50 yards apart, the lights being masked so that they only shone upwards. A flashing beacon was positioned 300 yards from the foot of the flare path to indicate the release point and entrance to the funnel – gliders were channeled from their release point down an imaginary funnel to pre-selected positions on the ground perimeter, otherwise the hordes of gliders landing at 30 second intervals presented a major collision risk.

The German technique was for a pathfinder aircraft to lay two groups of incendiary bombs, one just short of the dropping-zone and the other just beyond. It was a simple method that eliminated the risk of failure of faulty beacons, parachute pathfinders being injured on dropping or other mishaps. It provided pilots with a simple and unmistakable visual aid without greatly affecting the surprise factor, being launched at the last moment before the landings began. The Allies used a more sophisticated system; their air crews, trained up to Bomber Command standards, were considered to be more than capable of locating dropping-zones accurately in darkness. Those in command of the airborne forces believed that a simple landing plan with a whole brigade on to one dropping-zone, instead of separate battalion areas, would have advantages and that pathfinder soldiers should drop out of the main body to operate radio beacons.

DROPPING SUPPLIES

When airborne troops were isolated during a prolonged operation (such as Arnhem), Air Dispatch companies had to mount frequent air re-supply sorties flown by Allied aircrews, often at considerable cost to men and machines.

The Royal Army Service Corps was responsible for air cargoes dropped from aircraft, being divided into air and ground teams, the former responsible for loading and ejecting stores from aircraft, and the latter for jumping with the division and sorting out the stores on the ground.

By 1944 the Air Dispatch Companies handled three types of containers – the wicker pannier, the bombcell container, and the SEAC pack. The pannier weighed 500 lbs and was pushed into and despatched from aircraft by a roller conveyor; the hinged, two-compartment metal bombcells were just long enough to take the Enfield rifles or Bren-like machine guns and were slung on the underside of wings and the fuselage of any aircraft fitted with the universal bomb racks. Loaded with three or four bombcells, small fast fighter-bombers were particularly suitable for dodging the fire of anti-aircraft guns. The 200 lb SEAC pack was used in the Far East and many thousands were dropped into the jungles of Burma. On these supply missions various types and sizes of canopy were used, in Europe they were most commonly the 24 foot and 28 foot cotton-nylon or silk parachutes. Pilots took their aircraft down to a dropping level of 600 feet or below, depending on the type of parachute in use, and sometimes free-dropped non-breakable items at hedge-hopping heights of as low as 50 feet, without parachutes.

At Arnhem, it was recorded how the 1st Parachute Brigade jumped, to pattern the sky with brilliantly coloured parachutes and some 650 parapacks with bright yellow, red and brown 'chutes carrying guns, ammunition and equipment fell rapidly through the streams of para-troopers. Supply 'chutes, pushed out of the planes before the men jumped, floated down with numerous items of equipment such as miniature folding motorcycles.

The Germans used Versorgungsbomben (Provisions Bombs) – containers carried in bomb-racks of the aircraft and released in the same manner as bombs. About six feet long and 1½ feet in diameter, these containers held a parachute in a separate end compartment, the end-cap was torn off on release and the parachute pulled out. They had no shock-absorbers. Another type of container, four feet long and 18 inches square, used a simple static-line parachute at one end and a shock-absorber at the other.

German paratroopers dropped carrying sub-machineguns and pistols, rifles, heavier machineguns and ammunition being packed in separate containers (Fallschirmbombe), dropped with the men. In Denmark, Norway and Holland, separated from their containers paratroopers armed only with light hand weapons met with disaster, and at Crete men were needlessly lost attempting to recover equipment cannisters. Learning their lesson the hard way, it was decreed that men should jump with rifles strapped to their sides, although it was still necessary for heavier equipment and additional ammunition to be packed in separate containers.

AIRBORNE OPERATIONS IN EUROPE – WORLD WAR TWO

The airborne activities of both Germans and the Allies in Europe during World War Two make not only stimulating reading but, in most cases, are eminently suitable for reproduction on the war-games table, either as full-scale actions or Individual Skirmishes.

The German invasion of Crete and Operation 'MARKET GARDEN' at Arnhem, among the most extensive and best-known of airborne operations, are fully described and considered as war-games in these pages. But there are numerous other actions involving parachutists and glider-borne troops that are equally suitable for reproduction on table-top battlefields.

In Norway, Denmark and the Low Countries in April and May 1940 the Germans convincingly demonstrated the effectiveness of a combination of specialist airborne soldiers and ordinary infantry carried by air. Apart from the Battle of Britain, these operations did more to convince the British nation of the possibility of invasion than did the military disaster in France. Extremely successful and historically significant, these early airborne operations proved vital to the overall plan and achieved results unobtainable by any other means, using the surprise of an indirect and unexpected approach to soften the resistance of a startled enemy.

THE GERMAN INVASION OF NORWAY AND DENMARK

This, the first World War II airborne operation, was carried out on a grand scale with over 500 troop carriers – ten Gruppe of Ju52s and one Gruppe of Ju90 and Focke-Wulfe Fw200 transports. At Aalborg in Denmark, Hauptmann Walter's 1st Bn. 1st Parachute Regiment allowed one platoon of No. 4 Coy. twenty minutes to overcome opposition and secure TWO airfields; the main body of No. 4 Coy. had to capture intact a two-mile-long bridge near Copenhagen – the German paratroopers armed only with pistols overcame the startled Danes and had captured the bridge *before* collecting heavier weapons from containers. No. 3 Coy. set to capture Sola airfield near Stavanger in Norway, approached in Ju52's 30 feet above ground level, pulling up to 400 feet, when the men started jumping. They came down into a hail of machine-gun fire but providentially two Messerschmitt long-range Me110 fighters arrived over the airfield to provide a covering fire (General Student made support by Messerschmitts and Stukas an integral part of his doctrine). This was the first opposed parachute assault in history and owed its success to airforce fire-support – the artillery of airborne warfare. The paratroops hurled hand-grenades through the embrasures of two well-sited Norwegian defence positions on the airfield boundary, and within half an hour the Germans had taken their objective.

Less the two companies already mentioned, the 1st Bn. were to jump onto Fornebu airport outside Oslo and seize it as a point of entry for air-landed infantry. Visibility down to 20 yards forced the twenty-nine Ju52s carrying the paratroopers to turn back, but Hauptmann Wagner, commander of the follow-up aircraft carrying the infantry, ignored orders to return and went ahead with his landing. As the aircraft made its final approach, it was badly shot up by heavy machine-gun fire and many including Wagner were killed or wounded; the machine pulled up and the group headed back towards their base. Meanwhile the six Me11s flying over Fornebu to support the air-landing, planning to land on the captured airport when their fuel was exhausted, saw the other aircraft pull out and decided to land before runnning out of fuel.

All but one landed safely. The aircraft taxied toward each other and formed a group, turning their tails outwards towards the Norwegian defences so that their rear gunners could engage them with fire from their Mg 15 aircraft guns. but it proved unnecessary because, at the moment when the Germans were most vulnerable, the Norwegian defenders lost heart and abandoned their defences. The fighter-commander radioed base and the Ju52s returned to drop paratroopers who held the airfield until transports arrived and landed infantry.

British troops were arriving in Norway, and No. 1 Parachute Coy. was dropped to prevent the Norwegians from joining them at Andalsnes. Dropping dispersed and at too low an altitude, men were killed when parachutes did not open. This set off rumours in Britain that the Germans were dropping parachuteless men into deep snow-drifts! The surviving paratroops were rounded up by Norwegian forces, but caution had been forced on the defenders at a time when bold action was required. At Narvik, the British were firmly entrenched in the town and the Royal Navy commanded the sea approaches, so that the small German force could only be reinforced by air – but there was no airfield. Scraping together all available paratroopers, and some courageous soldiers from a mountain battalion who volunteered to jump without experience or training, the Germans fitted extra fuel tanks to Ju52s and dropped reinforcements in the area. Their great need for artillery was solved by sending Ju52s, heavily laden with arms and ammunition, to land on a frozen lake – unable to take off again, when the ice melted they sank to the bottom.

THE ASSAULT ON EBEN EMAEL

In the Spring of 1940, although hoping to maintain strict neutrality and conscious that Britain and France would aid if she was attacked, Belgium relied on a delaying defensive position protected by a forward line of outposts. However at Maastricht, where the nearness of the Albert Canal to the Dutch border made outposts impossible, reliance was placed upon the Albert Canal's deep cutting, its 100 yards width being spanned by three bridges

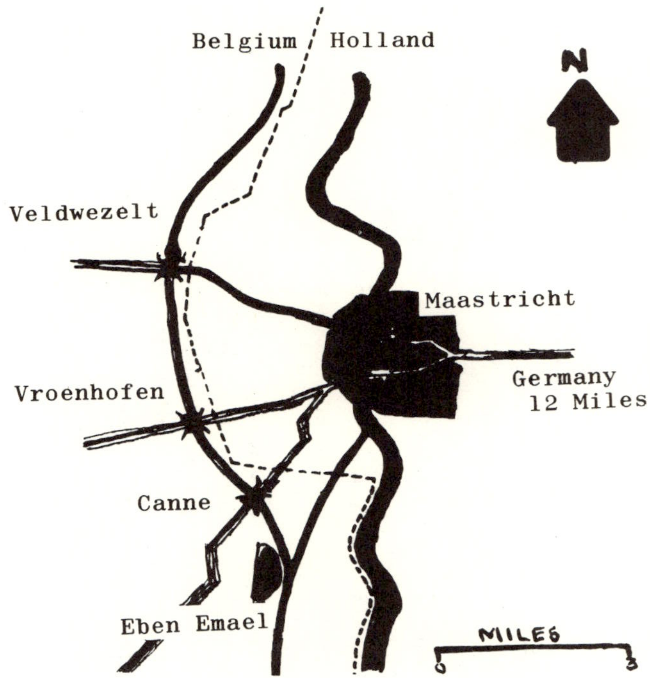

EBEN EMAEL AND CANAL BRIDGES

FORT EBEN EMAEL

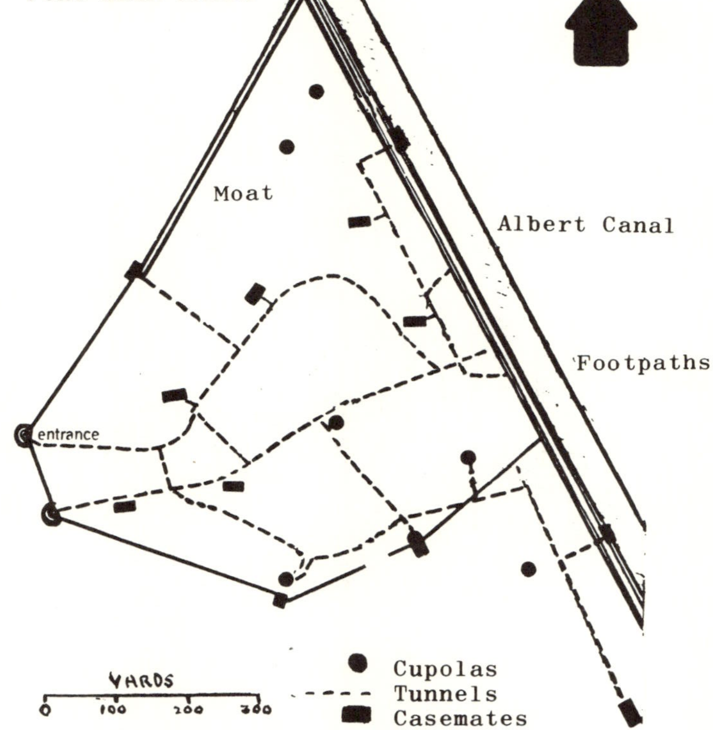

at Veldvezelt, Vroenhofn and Canne. The 7th Infantry Division were responsible for defending the area and had a brigade covering each bridge. The whole position was supported by the powerful artillery fort of Eben Emael.

Well prepared and sited, the bridge defences consisted of four massive concrete pillboxes on the near bank, one beside the road, mounted an anti-tank gun; another immediately below the bridge, and one on each flank some 500 yards distant, all mounted machine-guns. There was also a small post on the far (eastern) bank. The positions were garrisoned by a company positioned on the near bank at each bridge. Canne, the southernmost of the bridges and nearest to Eben Emael, had an anti-tank gun bunker set back into the hillside.

All the bridges had prepared demolition charges in position and could be quickly blown by demolition parties in the anti-tank bunkers. No Belgian forces operated east of the canal because of the nearness of the Dutch frontier but surprise seemed impossible because the Germans would have to fight their way across the 'Maastricht appendix' of Holland, so that by the time they reached the Belgium border the bridges would be demolished and well prepared defences confidently awaiting them.

With bitter memories of 1914 when the forts around Liege had been smashed into submission by heavy German siege-guns, in 1933/5 the Belgians had blasted out of natural rock the fortress of Eben Emael resembling the great defensive works of the French Maginot Line, with one side rising a sheer 120 feet from the canal. The other faces were protected by concrete pillboxes, 60 mm anti-tank guns, heavy and light machine-guns, ditches, a 20 foot wall, mine-fields and searchlights. The fort's armament consisted of six 120 mm guns in revolving armoured cupolas and eighteen 75 mm guns in cupolas or casemates, mounted in emplacements with walls and roofs of five feet thick reinforced concrete.

In November 1939, a special combat group under Hauptmann Walter Koch was formed from 1st Parachute Regiment 7th Air Division of Engineers, plus pilots, for the special task of capturing the three bridges intact and neutralising the fort. A parachute unit attacking in DFS 230 assault gliders to ensure a concentration of attackers on the objective, Sturmabteilung Koch trained intensively on fullsize mock-ups of the Eben Emael defence system until every man knew his own role and that of his comrades. Towed to the Dutch border the gliders were to be released at 8,000 feet, to glide silently across the Maastricht 'Appendix' and onto their objective, undetected by sound location devices of Belgium anti-aircraft defences.

Koch's assault force was in four distinct groups, each with specific duties: (1) Assault group 'CONCRETE' (Leutnant Schacht) to secure the bridge over the Albert Canal at Vroenhofen and hold until the arrival of ground forces; (2) Assault Group 'IRON' (Leutnant Schaechter) to secure the bridge at Canne; (3) Assault Group 'STEEL' (Oberleutnant Altman) to secure the bridge at Veldvezelt; (4) Assault Group 'GRANITE' (Oberleutnant Witzig) to land on the flat roof of the fort and cripple the artillery armament, holding on until the arrival of Army Engineer Battalion 51.

These trained and experienced engineers were armed with 110 lb hollow-charge explosives capable of punching a hole 12 ins in diameter through six feet of concrete.

Each towing a glider, thirty-one Junkers took off at 0430 hours 10 May 1940; once airborne they formed up and wheeled onto course in a steady thirty minute climb to just over 8,000 feet, moving towards the release point near the Dutch border, indicated by pre-laid ground beacons. Of all the gliders, that carrying Leutnant Witzig and part of his Eben Emael Group, broke a tow-rope and landed in a field deep in Germany; another of the same Group's gliders cast off too soon and never reached its objective. But the remainder reached the release point and let go tow-ropes in free flight. The gliders nosed gracefully down towards the Belgium frontier getting lower and lower until, one-by-one their skids touched down, they ran forward 20 yards and stopped.

Even before a wing tip touched ground their doors were off and soldiers poured out, as each bridge group (consisting of 5 Infantry and 4 Engineer sections) ran swiftly towards the pillboxes and bridges, returning fire opened upon them from the bridge defences. The advance proceeded in leaps and bounds until the

The Fort Eben Emael Wargame. A DFS230 assault glider lands by one of the turrets of the Fort which is immediately attacked by Witzig's group with explosive charges, grenades and flame-throwers.

Engineers were within striking distance of the bunkers, dramatically neutralised by having huge holes blown in their concrete by hollow-charges and then pouring flame and throwing grenades through the gaps. The two northern bridges at Veldvezelt and Vroenhofen were seized intact and demolition charges removed before they could be blown. The Belgian demolition firing party at Veldvezelt realised that the bridge was going to be captured and radioed to their H.Q., three miles north, for permission to blow but, disbelieving the story of a glider attack and the presence of German troops on the bridge, permission was refused.

At Canne surrounding hills delayed the gliders and they put down several hundred yards from their objective, so forfeiting surprise. The defenders put up a heavy fire that prevented the engineers storming the pillboxes; the bridge was controlled from Fort Eben Emael, itself under attack, so permission for it to be blown was readily given, and it went up in the face of its frustrated attackers.

At Eben Emael, the nine remaining gliders of Witzig's group landed with precision on the roof and disgorged attackers who ran at will all over the exterior of the fort systematically destroying the twelve emplacements from which fire could be brought to bear on the bridges and the surfaces of the forts. The paratroopers fired and flamed, threw grenades into the embrasures and loopholes, placed charges of TNT on turret-edges and gun barrels, jamming the turrets and destroying the guns; ventilators and periscopes were attacked. Hollow charges detonated on top of the emplacements and turrets blew holes through the armour and concrete, sending a jet of flame and molten metal into the turret structure to wreck internal machinery and kill or shock defenders. Those who survived were assailed by flame-throwers or small charges and grenades dropped through the holes. Blasting open steel doors, the invading paratroopers entered the fort and, once inside, were difficult to eject as the defenders had to attack up 60 feet of spiral staircase.

Crossing the Albert Canal in inflatable boats, Pioneers of an Engineer battalion brought heavy demolition charges, flame throwers, a Bangalore torpedo for wire cutting, and other materials for prising open the fort.

At 0830 a lone glider flew in from the East, across the Dutch and Belgian frontiers, circled over Eben Emael and touched down in one of the few remaining clear areas – it was Witzig, towed off by a relief-aircraft, arriving three hours late to take part in the assault.

At 0610 hours German aircraft, taking casualties from now fully alerted defences, dropped reinforcements and ammunition at each of the bridges. Weak Belgian counter-attacks against the captured bridges were repulsed, and the Canne Bridge group, strengthened and reorganised, cleared the last defenders from the demolished structure.

In a few hours, at a cost of six dead and 20 wounded, the 70 paratroopers had neutralised Eben Emael, so that when the main ground forces (who had been delayed outside Maastricht) arrived and assaulted the main entrance, the garrison capitulated. The successful German airborne assault has come to be regarded as a classic example of what can be achieved by a small force against powerful defences, exemplifying the fullest exploitation of all the advantages that lie with the attacker – here they included the virtual beginning of a war rather than a battle. Tactical surprise was achieved by the silent approach of the gliders in a never-before-demonstrated landing method against which no defensive tactics had been considered.

THE LOW COUNTRIES, MAY 1940

To their awed enemies, the German airborne operations in the Low Countries bestowed almost magical properties upon paratroopers. It was planned for the 7th Fliegerdivision, consisting of three parachute battalions and 16th Infantry Regiment, to make tactical parachute landings in a long carpet across the river barriers at Moerdijk Dordrecht and over the Nieuwe Maas at Rotterdam itself, and at the airfield at Waalhaven, to provide a 30 miles long route for the German Panzer forces across the water obstacles leading into Holland. If successful, the paratroopers, becoming weaker with the passage of time, would have to hold their objectives against increasingly strong Dutch counter-attacks. Should the Panzer divisions be delayed at the Dutch

'Grebbe-Peel' main defensive line running from the south shore of the Zuider Zee to the Belgium frontier near Weert, then the paratroopers could be overwhelmed. Simultaneously, General von Sponeck's 22nd Infantry Division (47th and 65th Infantry Regiments and a reinforced parachute battalion) was to land and seize the airfields around the Hague, then air-land infantry regiments to capture the Dutch Royal family and national leaders.

Preceded by heavy air attacks with Stukas, Dorniers Heinkels and Messerschmitts, wave after wave of Junkers 52s approached the various dropping-zones. Prager's Battalion jumped at the bridges near Moerdijk, companies landing north and south of the objectives and overpowered the defenders in a fierce but quickly settled action. Dutch artillery continually shelled these bridges and on the next day a mechanised group of the French 7th Army attacked the area but were driven off by Stuka attacks. By the end of that day, main German ground attacks were exerting so much pressure on the Dutch that the threat to the airborne troops diminished; von Hubicki's 9th Panzer Division reached the bridges on the morning of 12 May to find them still intact and in German hands.

The Germans were fortunate in that, although every bridge was prepared for demolition the Dutch considered them vital, believing that a bridge blown is lost for ever while a bridge in enemy hands may be recaptured.

In the built-up area of Dordrecht there was room for only one company to land; despite attacking strongly, the paratroopers could capture only part of their objective. Their commander von Brandis was killed when the Dutch counter-attacked and recaptured the railbridge.

At Waalhaven airfield south-west of Rotterdam, Schulz's 3rd Bn. 1st Parachute Regiment jumped immediately east of the field, where the defenders engaged them, then airborne infantry landed on the airfield and assailed the defenders in the rear, overwhelming them after a short but sharp action. Although under artillery fire, two-hundred and fifty landings were made on Waalhaven on the first day of its seizure and despite R.A.F. bombing the field remained operational throughout the battle. During the afternoon, two battalions 16th Infantry Regiment landed there then dispersed to take over Dutch military and civil vehicles and speed away to reinforce the isolated company fighting at the bridges and towns in Dordrecht, where resistance was maintained until 12 May. This action exemplified the limited offensive capability of lightly armed airborne troops lacking surprise against well defended objectives; Dordrecht was eventually captured by reinforcements strong enough to succeed without the benefit of surprise.

Von Sponeck's 22nd Division, with only one reinforced battalion of paratroops for the purpose, had three airfields to capture. Time was lost in locating the airfields on the featureless Dutch landscape so that Ju52s began to arrive before the fields were captured or cleared of anti-landing obstacles. Soon, destroyed or immobilised aircraft cluttered up the airfields so that further waves, finding no clear space to land had to turn back. At Ypenburg, eleven out of 13 Ju52s carrying the first assault company of 65th Infantry Regiment were shot down in flames by anti-aircraft fire. Von Sponeck's aircraft was damaged and had to put down in another field, but not before he had seen the airfields strewn with the wrecks of aircraft and general chaos on the ground. Subsequently Kesselring cancelled the attack on the Hague and ordered the 2,000 men of 22nd Division who had landed (another 5,000 were unable to join the battle) to head north of Rotterdam. Elsewhere, the Dutch defenders defeated the Division's attempt to establish spearheads and surviving Germans withdrew from the airfields during the afternoon.The landings of 22nd Division had been a tactical failure, serving as an example of the vulnerability of airborne troops when faced with alert and determined defenders. On the other hand, 7th Air Division were universally successful, ably demonstrating the tactical value of offensive air support to airborne warfare.

At the twin-bridges on the outskirts of Rotterdam twelve Heinkel 59 seaplanes carrying 120 men of Schrader's No. 11 Company 16th Infantry Regiment skimmed along the Nieuwe Maas until right alongside the bridges, when men swarmed ashore to cut demolition wires and take the bridge before the Dutch could recover. The Germans were reinforced by a Lieut. Kerfin's assault force of 50 paratroopers dropped into a sports stadium just south of the bridges, who hastened to the scene of the action in requisitioned trams. Dutch artillery fire made movement on the bridges impossible and the counter-attacks of now reinforced and determined Dutch soldiers managed to seal off the German bridgehead. In this built-up area unsuitable for tanks, attacks could make no progress and when the situation had remained static for 48 hours, Hitler ordered massive air attacks on Rotterdam which reduced the city to a charred shell. At this point Holland capitulated.

The Germans took heavy casualties in their airborne attacks on Holland; of the 2,000 men of 22nd Division who landed 40% of the officers and 28% other ranks were casualties; 170 Ju52s were destroyed and as many seriously damaged, with high air-crew losses that included many instructors borrowed from flying schools for the operation.

OPERATION COLOSSUS – THE FIRST BRITISH AIRBORNE RAID

In 1940, No. 2 Commando changed its name to 11 Special Air Service Battalion, with H.Q., one parachute and one glider wing. It was elements of this force who carried out Operation 'Colossus', the first British Operational airborne raid of the war. This initial test for the new airborne arm was the destruction of the Tragino Viaduct in the sparsely populated country south-east of Salerno, in the west coast province of Campagnia in Southern Italy. Although possessing limited aircraft and specialist airborne equipment, the 38 man force was of the highest quality, coming from the first British unit to be parachute-trained. The six converted Whitley bombers arrived over the dropping-zone at 21.30 hours on 10 February 1941; in the full moon dropping casualties were light, but two aircraft failed to drop their containers of arms and explosives. Nevertheless, there were sufficient explosives to destroy a main aqueduct pier and bridge which effectively breached the viaduct, the object of the operation. The force was scheduled to escape in three parties and rendezvous with a submarine, but attention was attracted to the area when one of the Whitley's crashed at the pick-up point on the coast. Reluctantly, the Admiralty recalled the submarine, stranding the paratroopers who were captured two days later. The operation had negligible direct military effect, but resulted in the Italians using troops to guard similar vulnerable targets – a 'fringe' benefit arising from the employment of airborne troops. Considerable experience was gained, particularly in dropping parachutists at night, leading to the selection of airborne troops for further commando-style operations.

OPERATION 'BITING' – THE BRUNEVAL RAID

In 1942, British scientists urgently required accurate information on the Wurzburg radar equipment that, from a chain of stations along the French Channel coast enabled the Germans increasingly to intercept and destroy R.A.F. aircraft. Aerial photographs and French Resistance Intelligence sources located one such station on the cliff edge near the village of Bruneval, 12 miles north of Le Havre – an ideal site for an airborne raid with sea evacution. Such a combined operation was planned for late February 1942, using C Coy. 2nd Parachute Bn. commanded by Major J. D. Frost, dropping from twelve Whitleys of 51 Squadron R.A.F., there was to be a naval evacuation force of assault landing-craft for the pick-up, support landing-craft, a motor gun-boat and two escort destroyers. Thirty-two officers and men from the Royal Fusiliers and South Wales Borderers formed a beach-protection party to come ashore from the support landing craft and cover the final withdrawal.

The radar apparatus was sited on the cliffs near a large villa where its technicians lived; a garrison of 100 infantrymen for the fifteen defence-posts along the cliffs lived in a farmhouse, La Presbytere, nestling in a wooded enclosure 400 yards along a track to the north. Another platoon, responsible for shore defence, lived in Bruneval village, to the south clustered on both sides of a ravine leading down to the beach. Behind the villa the ground was flat and open; there was an ideal dropping-zone 1,000 yards east between two roads which converged on the La Presbytere track.

The operational plan provided for three groups; dropping first the 40 men of 'Nelson' led by Lieut. Chartris were to knock out the machine-gun defences on the cliffs and then move into Bruneval village. 'Drake' force of 50 men divided into two sections, one under Lieut. P. A. Young to seize the Radar installation while Major Frost's party broke into the villa and sought prisoners. 'Rodney' group of 40 men under Lieut. Timothy, dropping last, were to prevent the Germans in La Presbytere from approaching the villa.

On 27 February, after three cancellations because of bad weather, the naval flotilla put to sea in the afternoon and the raiding party assembled at Thruxton airfield. The wind had dropped and visibility was perfect, although the French countryside was covered in snow. Approaching the French coast in bright moonlight, the twelve Whitleys came under fire from coastal anti-aircraft batteries and two of them had to alter course. Without exception, the drop was made with complete accuracy and 100 men came down in exactly the right area with only a few minor injuries. Flying in the two scattered Whitleys Lieut. Chartris and half the Nelson' group landed some 3,500 yards south of the villa, close to the village (and second objective) but a long way from the machine-gun defences on the cliffs, their main target.

Collecting their gear from containers, 'Drake' and 'Rodney' groups advanced in three parties across the open ground to the radar installation, the villa and the woods surrounding La Presbytere. Lieut. Young's party went straight to the huge Wurzburg Radar saucer and opened fire with their sub-machineguns, killing or dispersing the manning-crew. Throwing a cordon around the villa, Major Frost burst in with four men, rushing upstairs to kill a soldier firing at Young's party from a window. Frost now concentrated 'Drake' group around the Radar set, being rapidly dismantled by radio technician Flight-Sergeant Cox, R.A.F. aided by Captain Vernon's Engineers, which had almost been completed when the slowly reacting garrison at La Presbytere commenced firing on them from the edge of the wood, killing a paratrooper.

The withdrawal to the beach now began under fire from machinegun posts and pillboxes on the cliffs, still operative through the non-arrival of 'Nelson' group. Deep snow drifts hampered the party portering the radar equipment along a difficult path between snow drifts. News of the snow had reached the raiders at the last moment, and it had not been possible to obtain white camouflage clothing. Lieut. Timothy's 'Rodney' group had left the area of La Presbytere and were heading for the beach to take-over 'Nelson's task. Fortunately, at this critical stage Chartris and his delayed 'Nelson' group arrived, having been involved in a fire-fight with a German foot patrol near Bruneval. With great dash they fought alongside Lieut. Timothy's group, clearing the machine gun posts and pillboxes in line-of-sight of the beach.

Concentrated on the beach by 0215 hours, the raiding party could not contact the Naval flotilla by radio and sea-mist blanketed the Aldiss lamp, so they fired Very-lights that lit up the clear winter sky over the beach. Avoiding a German destroyer lying off-shore, three LACs and three MTBs made a determined line through the mist to the beach, where the raiders were taking casualties from heavy fire from the clifftop. The Royal Fusiliers and South Wales Borderers of the beach-protection party leapt ashore from the MTBs and with well-aimed automatic and rifle fire at the clifftops contributed to the racket that disturbed the crisp night air, as the wounded and the captured equipment was loaded on the LACs. The last infantryman left the beach and, once out to sea, the raiding party were transferred to the motor-gunboats and the force headed for home, a squadron of Spitfires swooping down to escort the flotilla to Portsmouth harbour as dawn broke. At a cost of three men killed and seven wounded, the raid had secured much vital equipment, with a captured German Radar expert who provided more information.

GERMAN AIRBORNE ATTACK ON THE CORINTH CANAL BRIDGE

On 26 April 1941, in a belated attempt to take the Corinth Canal bridge and prevent the Allies from pulling back to evacuation ports in the Peloponnese, Oberst Sturm's 2nd Regiment supplied 1st and 2nd Bns., plus a parachute artillery battery, engineers, signals and medical teams in a single lift. Landing immediately before the main drop, Junkers 52s and three DFS 230 gliders took in an engineer party to prevent the bridge over the Corinth Canal being blown after evacuting troops had crossed.

Several British infantry companies were deployed in defence of the bridge, with a few light tanks in support, and some 40 mm anti-aircraft guns mounted around the actual crossing itself.

Under cover of an apparent routine air attack, the glider-tug combinations approached the bridge, cast off and landed accurately near both ends of the objective, getting behind the local defences which were overwhelmed and engineers began seeking demolition wires. The parachute aircraft dropped 1st Bn. Group north of the canal and 2nd Bn. Group to the south, descending almost onto the heads of the confused defenders. The German 1st Bn. triumphed relatively quickly and easily, but 2nd Bn. were fiercely opposed by defenders dug in south of the canal, but by noon the defenders were overcome.

On the bridge, the assault engineers dismantled the explosive charges and piled them neatly in the centre of the road. Then a shell from a Bofors anti-aircraft gun (being used in a field-artillery role to engage the Germans on the bridge) fortuitously struck the pile of explosives, setting them off with a great roar. The iron bridge structure rose into the air, then its wreckage fell into the canal, effectively blocking it. The surviving engineers made an improvised crossing that was used by 5th Panzer Division on the following night.

A tactical success, the operation cost 240 casualties – a small price to pay for cutting-off and later capturing some 10,000 British Commonwealth, Greek and Yugoslavian soldiers.

OPERATION 'FRESHMAN'

An ill-fated glider-borne attempt by volunteer airborne engineers to destroy a German heavy-water plant situated in a deep valley with thickly forested sides at Vemork in Norway. The dropping-zone was some five hours march away from the target; landing signals were by R.A.F. Eureka signal-device planted in the area by Norwegian agents. On 17 November 1942 two Halifax bombers towing Horsa gliders took off from Scotland on a north-easterly course over the North Sea, crossing the Norwegian coast on correct bearings before flying into a blizzard.

In one Halifax tug, the Rebecca radio-unit failed to make contact with the Eureka ground-beacon and the pilot, realising he had insufficient petrol to search for the landing-zone turned for home. As he did so, the tow-rope snapped and the glider crash-landed in thick snow on a mountainside, killing three men on impact. The survivors were quickly rounded up by German troops and executed by firing-squad within a few hours. The Halifax hit a mountain-top, killing the entire crew.

The second combination made a successful landfall but the glider cast-off prematurely and diving blindly through dense cloud, smashed into high ground, killing eight and injuring four of its 17 occupants. The remaining five men were captured before they could get away from the area, and were later shot by the Gestapo, who also poisoned the wounded men in hospital. The Halifax returned safely to base.

ALLIED AIRBORNE OPERATIONS IN NORTH AFRICA NOVEMBER/DECEMBER 1942

The American 2nd/503rd Parachute Infantry Bn. and the British 1st Parachute Brigade were included in Operation 'TORCH' order-of-battle, to carry out raids or operations in close support of their respective national forces. U.S. Army Air Corps C47s were to be used as carrier-aircraft and, as far as possible 1st British Parachute Brigade were given a 'crash' conversion course in jumping from this aircraft, but a high proportion went to Africa without experience of the aircraft.

On the night 7/8 November 1942, U.S. 2nd/503rd took off from England in thirty-nine C47s for a 1,500 mile journey to occupy La Senia and Tafaraoni air bases. During the night flight the guide plane lost its way and actually landed to ask a French-speaking Arab the general direction of Oran! The force, running short of fuel, made an unopposed landing at Sebkra; in the event, U.S. ground forces had already taken the objectives before the airborne group arrived.

British 3rd Parachute Battalion flew to Maison Blanche airfield twelve miles from Algiers – the airborne contingent consisted of two companies, the mortar platoon, HQ Coy and a field ambulance. The remainder of 1st Parachute Brigade – 1st and 2nd Bns. and the balance of 3rd Bn. – arrived in Algiers by sea and reached Maison Blanche eleven days later, where they were joined by 2nd/503rd U.S. Parachute Battalion. On the same day, 13 November, the airborne contingent had dropped and occupied Bone Airfield without opposition, losing four men through accidents, unaware that Junker aircraft carrying Fallschirmjager ordered to drop on the same airfield had observed the British drop and flown back to Tunis.

On 14 November, the Americans dropped to seize and hold air fields at Tebessa and Youks les Bains, and 1st British Bn., due to drop near Beja and secure the Souk el Arba plain, were prevented from dropping by thick fog. They took off on the following day with orders that no weather conditions were to prevent a second landing. Dropping unopposed, they found their main obstacle to be local Arabs who were attracted by the contents of the weapons containers and also by the silk parachutes. The parachutists commandeered a fleet of old buses and made contact with French Native troops with whom they attacked an Italian tank leaguer, when their commanding officer Lt.Col. Hill was wounded. Subsequently, the battalion fought as ground troops, conducting guerilla operations and ambushing enemy columns.

John Frost's uncommitted 2nd Para. Bn. were to be dropped on Depienne airfield, south of Tunis to destroy enemy aircraft and stores; then to move westward to meet the advancing 1st Army. The U.S.A.C. provided forty-four C47s and on 29 November the battalion were dropped in ploughed fields some 12 miles from the target area. Marching by night, the battalion reached the airfield on the following day but found neither aircraft nor stores, so they pushed on to Oudna airfield 18 miles from Tunis but found that the airfield was not being used. With the Allied advance halted, they were trapped 50 miles behind enemy lines and had to make a fighting withdrawal. Before leaving, the parachutists were attacked by German tanks supported by Me109s, later a flight of Ju87s sought in vain to discover their well-concealed positions around the airfield.

Leaving a small party to protect the wounded from hostile Arabs, the battalion moved off and came under attack before going very far. B and C Coys., with A Coy. in reserve, set-up a defensive position based on two small hills, to fight off German lorry-borne infantry, supported by tanks, heavy machine-gun, mortar and artillery fire. C Coy. were virtually wiped out, and when the retreat was resumed, the medical officer remained behind to look after the large numbers of wounded. Pursued all the way, the survivors of the battalion moved slowly over hilly paths, ploughed fields, rivers and ditches, being given food by friendly French farmers who warned them against treacherous Arabs. Dispirited by increasing numbers of casualties, the parachutists were relieved to meet an American scout-car patrol and, on the evening of 3 December, 180 men – less than 25% of the full battalion strength –reached Allied positions at Medjez.

This was the end of Allied airborne operations in North Africa, partly because of inclement weather but mainly because 1st Parachute Brigade were needed in the line as infantry, where they achieved a reputation as an elite formation known to respectful opponents as 'Roten Teufeln' (Red Devils). In the North African campaign, the British Parachute Regiment lost 1,700 men killed, wounded and missing.

THE GERMAN ASSAULT ON LEROS

Leros was one of the Dodecanese Islands occupied during airborne raids by the 11th Para. Bn. and the Iraqi Para. Coy., raised and trained by the Royal Air Force. It was garrisoned by some 3,000 British troops supported by Italians and Greeks but, being beyond the range of Allied fighters, daylight naval aid was not possible. The Italians and Greeks were in coastal defence positions with the British in reserve when on 12 November 1943, after preparatory air attacks, the Germans made two successful beach landings. While Allied counter-attacks were being delayed by the Luftwaffe, in a manner resembling old times, a formation of Junker 52's, in their last-ever daylight flight over Allied territory, dropped a battalion of 2nd German Parachute Regiment onto the mile-wide neck of land between two bays. This close support of the seaborne assault paralysed the defenders when they should have been concentrating on throwing back the attack from the sea. The paratroopers cut the island in two, causing a confused situation that made it difficult for the British commander to decide who best to counter-attack. Finally he threw his reserves against the dug-in paratroopers, who held them off while reinforced seaborne troops expanded their beachheads and eventually forced an Allied capitulation.

THE INVASION OF SICILY – OPERATION HUSKY

The Allied seaborne assault on Sicily was to be headed by the first large-scale Allied airborne assault, by units of 1st British and 82nd U.S. Airborne divisions. In number and in quality both were quite capable of carrying out their allocated tasks, although Air Force and glider units lacked the training and experience to adequately carry out such operations in darkness. General Hopkinson, commander 1st British Airborne Division, had persuaded General Montgomery to begin the assault with a massed NIGHT glider descent in difficult country by inexperienced crews flying unfamiliar American gliders.

The 1st British Airborne Division was composed of Brigadier Lathbury's experienced 1st Brigade; Brigadier Down's 2nd Brigade, and the 1st Air Landing Brigade of Brigadier Hicks, consisting of two infantry battalions – the South Staffords and the Border Regiment. The Division was brought up to full strength with Brigadier Hackett's 4th Parachute Brigade from the Middle East. General Ridgway's 82nd U.S. Division was formed of Colonel Tucker's 504th and Colonel Gavin's 505th Parachute Infantry Regiments and the 325th Glider Infantry Regiment.

XII U.S. Troop Carrier Command provided 331 C47s; 222 being allocated to the 82nd Division and 109 to the 1st British Division, who also had a squadron of 28 Albemarles and seven Halifax's, supplied by 38 Wing R.A.F.

From America 140 Waco gliders arrived in crates, to be assembled by gliderpilots, like making up a model-kit. Lt.-Colonel Chatterton, British glider pilot commander, ferried giant Horsa

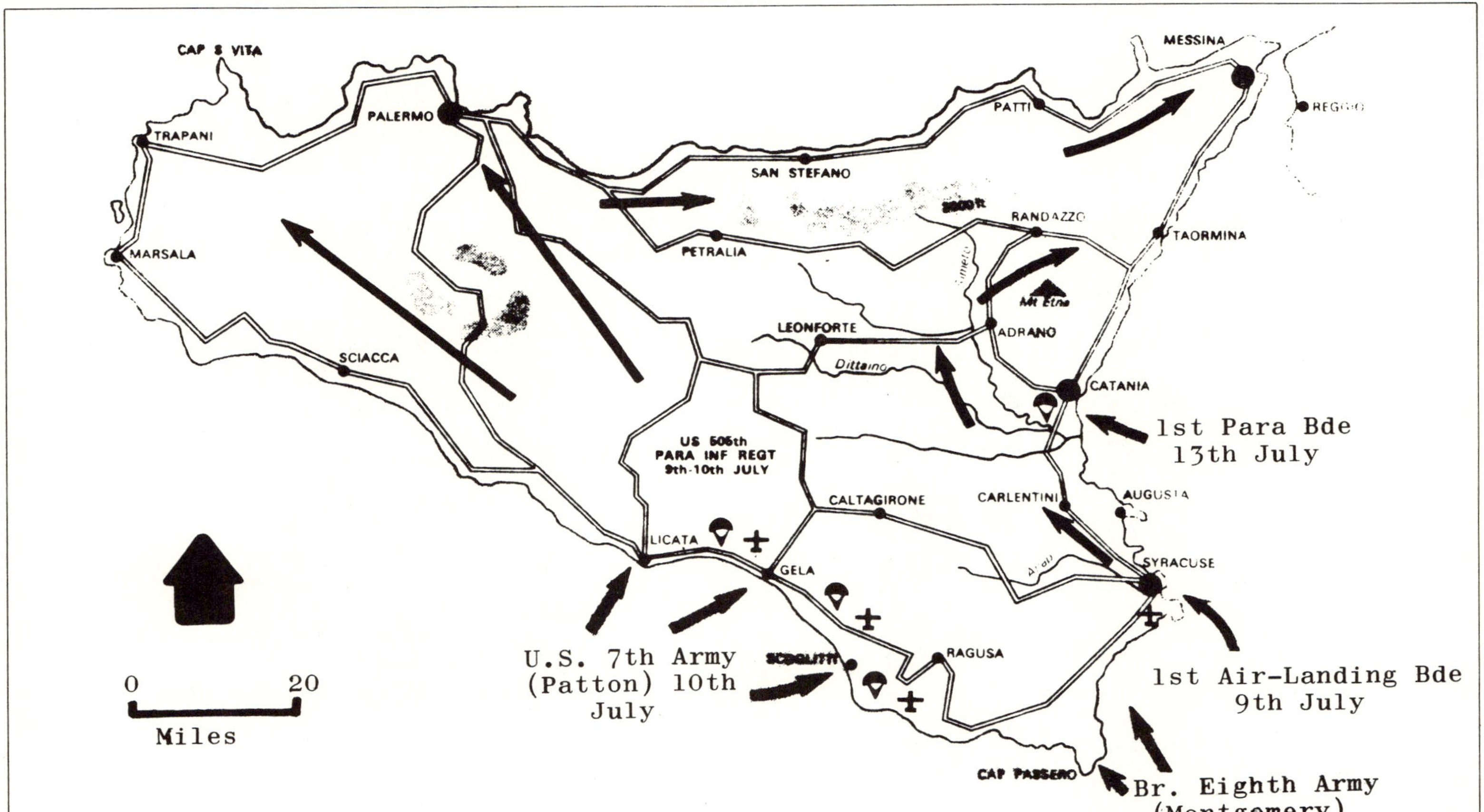

THE BRITISH AND AMERICAN LANDINGS IN SICILY

gliders, big enough to carry anti-tank guns and their towing-jeeps, from England to North Africa; nineteen arriving out of the twenty-nine that set out.

Supporting their own countrymen of 1st U.S. Infantry Division, Gavin's 505th Regiment were to drop inland of the invasion beaches and secure Piano Lupo, vital high ground in the Gela area, plus other tasks. Their C47s, in 'V' formations of nine aircraft, were to fly just above sea level and rise to 600 feet during the final approach with pilots identifying D.Z.'s from aerial photographs; pathfinders were not used and the zones not marked.

Hicks's Air Landing Brigade were to capture the important Ponte Grande bridge near Syracuse in a three phase attack – at 2315 hours 9 July, two Coys. in eight Horsa gliders were to land close to the bridge and seize it. At 0115 hours 10 July, the main body of the brigade in 136 Waco gliders were to come down on a larger L.Z. some two miles away and then move to the bridge, detaching one company to deal with a coastal defence battery. Finally, the South Staffords were to hold the bridge while the Border Regiment pressed on into Syracuse. Landing from the sea further south, ground troops were to link up by 1000 hours, 10 July.

Flying individually at low level from North Africa, gliders were taken up to 1,900 feet before being released 3,000 yards out to sea, to glide inland while the tug-aircraft returned. It was anticipated that all gliders would land within a 20 minute period.

Two thousand British troops of the Air Landing Brigade in 137 American Waco and eight British Horsa gliders took off from Tunisian airfields on the evening of D−1 (9 July). Seven gliders did not get as far as the North African coast and about 90% of the tugs entered the second leg of the journey from Malta with their gliders still in tow. The wind had increased to gale proportions and conditions were worsened by flying low to avoid radar detection, and coast landmarks were blotted out by a wall of dust raised by an off-shore wind. There was but sparse moonlight.

These factors, perhaps accompanied by timidity on the part of the inexperienced pilots of the tug-aircraft, caused a majority of them to turn too soon for home so that about 60% of the gliders were prematurely parted from their tugs because glider pilots had to blindly slip their tow-ropes. Seventy-five Wacos and three Horsas crash-landed in the sea, a few of their occupants being picked up from the floating wooden wreckage by passing assault-craft, others including the Brigade Commander swam for the shore. At least 252 men were drowned.

Only 52 gliders made landfall, a mere twelve landing anywhere near the target, the rest being widely scattered over an area covering 25 miles of coast. Of the six Horsas due to land beside the Ponte Grande, three landed two miles distant and their troops reached their objective later, and only two came down in the area, one of them hit the bank of the canal at speed and an explosive device destroyed aircraft and passengers, the other Horsa landed accurately and intact, following the beam of a convenient searchlight right down to the ground. It contained a platoon of 14 men of the South Staffs under Lieutenant Withers, who sent seven men to swim across the river and then the two small parties attacked from north and south to capture the bridge and remove the explosives. During the night small groups joined them until at first light there was a mixed force of seven officers and 80 other ranks of the South Staffs and Borders; they were armed with their personal weapons, plus one 3-inch and a 2-inch mortar and four Bren guns. Constantly under shell-fire and attack by Italian infantry, the force held on until only 15 men remained unwounded. Running out of ammunition, at 1500 hours they were overrun, but the Italians could not destroy the bridge as the explosive charges had been removed. British infantry eventually arrived at 1615 hours (more than six hours late), mounted an attack and secured the bridge. Hick's Air Landing Brigade suffered 490 casualties, plus 88 casualties among the glider-pilots.

The aircraft carrying the American 505th Regiment lost formation and missed check points in the darkness, losing direction so that pilots approached Sicily from all points of the compass. Eventually, Gavin's Regiment were dropped some 30 miles from their correct dropping zone, scattered between Gela and Modica, some up to 65 miles off-course. Although numerous groups and individual paratroopers fought courageous and successful actions against the enemy wherever they encountered them, less than 200 men out of the 3,400 dropped were on the important high ground of Piano Lupo.

On the second night of the invasion, the 504th U.S. Parachute Regiment were flown in as reinforcements to jump over the American-held Gela-Farella airstrip. All naval and army commanders and formations were given strict orders that anti-aircraft gunners were to hold their fire while the C47's were overhead, and all was quiet when the first troop carrier formation arrived and their drop went smoothly. But, as following formations approached the dropping-zone, a single nervous machine-gunner set off a contagious outburst of anti-aircraft fire and within seconds every army and naval anti-aircraft gun for miles around was blasting away. Twenty-three of 144 aircraft that left Tunisia were destroyed, 37 were badly damaged and eight returned with their passengers; 229 of the 2,000 paratroopers became casualties during the night – some were shot after landing when ground units convinced themselves they were facing a German airborne attack. Subsequently, the troop carriers dropped their men as inaccurately as on the previous night and by late afternoon on 12 July, only 558 of the 504th remained, although they had hardly been involved in ground-combat and had flown over no enemy held territory.

To secure 8th Army's line of advance to Catania, 1st Bn. Parachute Bde. set out on 13 July to attack Primsole bridge spanning the River Simeto a few miles south of Catania. They emplaned in 105 C47's and eleven Albemarles with Halifax's and Stirlings towing eight Waco gliders, and eleven Horsa's carrying twelve medium a/tank guns and crews, engineers and a medical section. Lathbury's Brigade (1st, 2nd and 3rd Bns. – 1,900 men) were to land on four DZ's and two glider LZ's, west of the main road within two and a half miles of the bridge. The 1st Bn. were to approach the bridge from both sides and secure it; the 2nd and 3rd were to take the high ground south of the Gornalunga Canal and north of the Simeto river. It was to be the first occasion when 21st Independent Parachute Company acted as pathfinders to light the glider LZs.The 50th Division and an armoured brigade were expected to link-up with the paratroops during the following morning.

1st Airborne's R.A.F. Advisor prevailed upon the American C47 pilots to abandon their usual formation tactics and fly as a 'bomber stream' (pairs of aircraft separated by one and a half minute intervals) without appreciating that these aircrews, inexperienced in night navigation, relied on following-their-leader. Then Allied naval anti-aircraft gunners brought down two aircraft and damaged nine that had to turn back. This caused even greater dispersal and loss of course, accentuated by Axis Anti-aircraft fire that shot-down another 37 aircraft and forced ten more to drop out and return home. Taking violent evasive action, pilots haphazardly dropped paratroopers and cast off gliders, thus 39 aircraft dropped their troops within half-mile of DZs but the remaining 48 pilots ditched them many miles distant. Four out of the eight gliders that landed intact were in the right place, nine others crashed on landing or were lost at sea.

At dawn, of the 1,900 men who had taken off from North Africa, only 250 from 1st and 3rd Bns. with three anti-tank guns were on or around the bridge. The bridge demolition-charges removed, Lt. Col. Pearson (1st Bn.) in command of bridge defences, ordered his troops to dig-in on the north side of the river; three anti-tank guns, two 3-inch mortars, light machine-guns and a Vicker's were sited and the road mined.

On the previous day German General Heidrich's 1st Parachute Division had dropped south of Catania to reinforce the defenders – excellently demonstrating the value of airborne reserves during defensive operations. These experienced high-class troops were now sent against the British position at the bridge. Throughout the day Pearson's force resisted strongly against increasingly heavy German attacks by paratroopers, then infantry supported by tanks and self-propelled guns; fighter aircraft strafed incessantly. By early evening the survivors were forced to withdraw to south of the river but still denying the bridge to the enemy then, under cover of darkness, they moved south to link up with Frost's 2nd Battalion operating in the hills south of the river. Next day contact was made with 50 Div. infantry and tanks who had fought their way to a hill a mile short of 2nd Parachute Bn.'s position, but it was not until the 15th that the infantry and armour were able to attack the bridge in what was a costly and unsuccessful assault. Several forceful attacks by British tanks, paratroopers and infantry were held by the Axis troops until, during the night of the 16th, Pearson led an attack by an indirect route and recaptured the intact Primosole bridge before dawn.

In Sicily, the 1st Airborne Division lost 454 dead, including 57 glider pilots, 240 wounded and 102 missing.

The Allied High Command were well aware that the airborne operations in Sicily (described by General Gavin as 'a self-adjusting foul-up') had been so technically inadequate that they put in doubt the whole future of airborne warfare. A painful Pyrrhic victory through inadequate crew training, errors in airborne methods and tactics, shortage of air transport and adverse weather conditions, nevertheless it is doubtful if without that experience faults would have been righted in time for D-Day.

AIRBORNE OPERATIONS IN ITALY

On 12 September 1943, when German counter-attacks were threatening to throw the Allies back into the sea at Salerno, at six hour's notice 504th U.S. Para. Inf. Regt. was dropped as reinforcements immediately behind the Allied Lines. Joined twenty-four hours later by 505th U.S. Para. Inf. Regt., they were used in the same way as Heidrich's Fallschrimjager in Sicily when they helped to turn the tide of battle.

At another stage of the Salerno operation, 2nd Bn. 509 U.S. Para. Inf. Regt. were dropped by night to block roads in the area of the Italian village of Avellino some 20 miles from the beachhead. Surrounding mountains forced the drop to be made at a great height, accurate identification of release points in the darkness was difficult and there were no proper D.Z.'s so that men were scattered over 100 square miles and completely failed to carry out their task. More by luck than judgement, 510 of the 640 men who jumped eventually returned to Allied lines.

Operation 'Hasty'

To speed up the advance on Rome, General Leese, commanding British 8th Army, had 2nd Independent Parachute Brigade dropped behind the German lines to prevent delaying demolitions. Just before daylight on 1 June 1944, three officers and 57 men were dropped from three C47s; 'dummy paratroops' were also dropped to confuse the enemy. Ordered to 'occupy the dominating heights by day and descending to the valleys and wreak havoc by night', in three groups they harassed road communications for a week, spreading apprehension but achieving nothing of military importance. About 50% filtered back to their own lines a week later, after an operation pointing the lessons that during the early stages of its mission an airborne force can often achieve results out of all proportion to its size but, if early evacuation or relief is not possible, then the situation becomes increasingly difficult and survival is balanced on a narrow edge between fighting it out or evading contact with the enemy.

Operation 'Eiche' – the German airborne rescue of Mussolini

In late 1941 General Student detailed Oberstleutnant von der Heydte to raise and command the Parachute Lehr Battalion, a unit to develop techniques of paratroopers firing weapons during descent; night dropping and assembly; parachuting from a converted Heinkel III bomber, and 'diving-glider' approach.

On 12 September 1943 this force carried out a remarkable raid to rescue Mussolini from his Italian captors in the Hotel Camp d'Imperatore on the Grand Sasso, a remote ski-hotel perched on a narrow ledge surrounded by mountains on three sides and by a precipitous cliff falling away to the valley below, from which its only access was by a funicular. The hotel and lower funicular station was guarded by Italian Carabiniere units.

Terrain ruled out approach by land or parachute, so a company of the Parachute Lehr Battalion, reinforced by a team of SS Commandos, were to land on the ledge in twelve Fs 230 gliders, while the remainder of Major Mors battalion moved by road to seize the funicular station at the bottom of the valley. Towed by Henschel 126s' the tug-combinations took off from Rome and, in spite of landing conditions being worse than anticipated, the gliders touched down successfully. Covered by the men in the first two gliders, SS Hauptsturmfuhrer Skorzeny and his team landed and quickly seized the hotel, the astonished Italian guards passively surrendering. In the valley below, the funicular station fell after a little fighting.

Piloted by Hauptmann Gerlach, a Fiesleler-Storch landed on the shelf, and the horrified pilot was told that he had to take off from the short strip with both Mussolini and Skorzeny as passengers! Laden with three men, the little aircraft's engine roared up to full power, soldiers grimly hanging on the wings and tail to prevent it moving until Gerlach signalled. As they let go, the machine surged forward, overloaded in the thin mountain air, when it reached the end of the strip its speed was insufficient for flight and although its nose lifted, the aircraft fell back going out of control as it struck its port wing on the rocks. Veering left, the small aircraft shot off the edge of the precipice and dived into the valley, gathering flying speed that allowed the pilot to recover full control. Despite a damaged wheel, the aircraft made a two-point landing at Rome to successfully complete a daring and skilfully conducted small operation.

The difficult task of conducting the mission was assigned to Major Mors, although SS-Hauptsturmfuhrer Otto Skorzeny later assumed all credit for the rescue.

Airborne Operations on D-Day, 6 June 1944

Prior to D-Day, there were four Airborne Divisions in Britain – the veteran 1st British and 82nd U.S.; Gale's newly formed 6th British; Taylor's 101st U.S.; and the Polish Parachute Brigade. IX U.S. Troop Carrier Command had 1,166 C47's; R.A.F. 150 C47's, 88 Albemarles, 80 Stirlings, 36 Halifax's; and between them the Allies could muster 2,591 gliders. In all, an airlift sufficient for two complete divisions.

The American 82nd and 101st were to seal off the Cotentin Peninsular to prevent reinforcements coming up from the South – 82nd dropping north-west of Carentan to capture the Douvre River bridge; 101st to land just inland from Utah beach north of Carentan; seize the western ends of the causeways leading inland from the assault beaches; secure crossings over the Merderet River, then advance westwards, link up with the 82nd and seal off the whole Peninsular. As these operations were considered more important the American formations were allocated sufficient aircraft for each to land three regiments plus support units in the initial wave.

The 6th British Airborne Division could land two out of its three brigades, plus supporting groups. Their objectives were to secure bridges over the Caen Canal and Orne River (marking the left flank of the Allied attack); to neutralise the coastal battery at Merville before the assault fleet came within its range; to destroy four bridges over the River Dives and as far as possible to dominate the area between the Orne and the Dives.

The three Divisional Commanders were well suited for their role – the Americans Mathew Ridgway and Maxwell Taylor were both veterans of Sicily and Italy and Richard Gale, although not previously serving with airborne forces, seemed to have a particularly clear understanding of the relative merits of gliders and parachutes, besides being a highly competent and experienced soldier. Cross-postings from 1st Airborne had given the 6th some battle-experienced leaders, such as Brigadier Hill of 3rd Parachute Brigade, Lt.Col. Pine-Coffin commanding 7th and Lt.Col. Pearson commanding 8th Para. Bns.

Air Chief Marshal Sir Trafford Leigh-Mallory, in command of air operations, believed that enormous casualities would be incurred and that the drops would result in 'futile slaughter'. His pessimistic views were overridden by General Eisenhower, the Supreme Commander.

The airborne forces were to be carried in their own national aircraft and standard operating procedures had been adopted by Allied Troop-Carrier Squadrons, who were capable of carrying out highly efficient daylight parachute and glider tasks but lacked experience and training in night navigation, other than onto clearly marked dropping-zones. In the event the performance of American crews resulted in an ugly crisis of confidence within the U.S.Airborne forces as to their chances of being dropped accurately into battle by their troop-carriers.

Air photos revealed areas of Normandy planted with anti-landing obstacles, ten feet poles probably linked with mines that became known as 'Rommelspargel' – Rommel's asparagus. Plans were amended to avoid the worst areas; para-drops were substituted for glider landings in the British zone – the gliders to come in on Lift Two, when the Engineers had cleared the area.

The British Airborne brigades were given these tasks – Poett's 5th were to capture the bridges over the Caen Canal, and the River Orne at Benouville and Ranville. A party was to land in gliders five hours before dawn, then paratroops were to drop in brigade strength to take over the bridges, and establish defensive positions in the surrounding villages, orchards and farmland. At the same time, Hill's 3rd Brigade were to neutralise the heavy calibre guns in the concrete-emplaced coastal battery at Merville, before daylight exposed the invasion fleet. Then, the bridges at Varaville, Robehomme, Bures and Troarn were to be destroyed to prevent enemy reinforcements passing south; the ridge Le-Plein-Le Mesnil-Troarn dominating the Dives river-line and the east-west lines of communication, had to be seized, before the brigade concentrated in the woods and orchards of the Bois de Bavent. On the afternoon of D-Day, Kindersley's 6th Air Landing Brigade were to be brought in near Ranville to aid in securing the canal and river crossings.

At 2303 hours 5 June 1944, six Albemarles carrying 60 pathfinders of 22nd Independent Para. Coy. took off as the spearhead of the invasion of Europe. Immediately following were six gliders bearing the Oxford and Bucks Light and Royal Engineers,bound for the Caen Canal and Orne bridges; then the two Parachute Brigades were airborne, followed by an assault party from 9th Para.Bn. in the three gliders that were to make a pin-point crash-landing on the Merville battery.

The night sky pulsated with the throb of more than a thousand aircraft engines as the Allied airborne troops began the invasion of France on a night of grey cloud that screened the moon, as light rain misted the cockpit windows.

When 82nd and 101st U.S. Airborne Divisions began dropping their leading units at 0130 hours on D-Day it became immediately apparent that their pathfinders had landed wide of the mark, frequently failing to locate the exact DZs. This caused 101st Division to be scattered over a 15/25 miles area, with 1,500 paratroopers either killed or captured immediately after landing, and about 60% of their equipment lost when bundles were dropped into swamps and woods. Enraged paratroopers reported that the pilots, hedge-hopping to avoid flak, were flying too low for parachutes to open. Most of the American difficulties were caused by deficiencies in air-crew training and performance, and poor weather; the night glider landings were made at a high cost to life and stores – only a small proportion of anti- tank guns and equipment reached the paratroopers who badly needed them. Daylight reinforcement-landings were much more successful.

In spite of concentration difficulties, 101st quickly seized objectives behind Utah beach and when the glider regiment was flown in later, the western ends of the causeway leading inland and the bridges over the Douvre River were quickly taken.

Hindered by dense clouds, the 82nd Division drop was widely dispersed and they were forced to fight isolated battles, failing to occupy the river's banks, or to destroy its bridges. Rendezvousing on the village of St. Mere- Eglise, as instructed was almost the only part of their plans that went right. But by courage and determination they overcame all difficulties, and by dawn were holding a firm divisional perimeter around the area, and west of the Merderet River were fiercely engaged in preventing the German 91st Division from moving eastwards. Although handicapped by the death of their commander, throughout the day this Division made determined but unsuccessful efforts to dislodge the American paratroopers and link up with their isolated beach defences.

Despite their scattered delivery, both American Airborne Divisions succeeded in most essentials without the predicted fearful casualties from anti-aircraft fire and enemy fighters. At D-Day plus one, both divisions had linked up to form a six-mile bridgehead.

On DZ 'N' near Ranville, as some 2,000 men of 5th Parachute Brigade disentangled themselves from their parachutes, they encountered some opposition from German defences alerted by the pathfinders. Men lost their bearings in the waist-high Normandy corn, and by 0300 hours Lt.Col. Pine-Coffin was accompanied by only about 40% of his 7th Para. Bn. as he moved off to link with Howard's men on the river and canal bridges. Behind them, airborne engineers worked hard clearing the area of obstacles for the glider landings that were to come later in the day.

The airborne sector presented an incongruous and colourful scene at daybreak with hundreds of coloured silken canopies strewn in the fields and draped across hedgerows; parachute harnesses hung limply by rigging-lines from the branches of trees, and empty containers were scattered across the dropping-zones; some lay unopened in lanes and ditches.

Johnston's 12th Bn. of the brigade captured the village of Le Bas de Ranville and dug themselves in, as Luard's 13th Bn. occupied Ranville le Mariquet. Both were taken in face of fierce resistance and held against counter-attacks supported by tanks and self-propelled guns. In spite of the Rommelspargel, some cargo gliders had been included in this force and the anti-tank guns and crews were invaluable, destroying at least one tank and three self-propelled guns.

By dawn, B Coy. 7th Para. Bn. had reached the Caen Canal Bridge and came under heavy counter-attack; the battalion, occupying Benouville, withstood eight counter-attacks and many attempts to infiltrate its defensive positions. At 1000 hours General Gale and his Staff, wearing red berets, arrived on the scene; at 1300 hours, five hours before other seaborne units, Lord Lovat and a lustily blowing bagpiper marched in at the head of No. 1 Commando, to link with the paratroopers in holding their objective.

Six gliders released their tow-ropes 5,000 feet above the mouth of the Orne and divided, to aim for the canal bridge L.Z.'X' and the river-bridge L.Z. 'Y'. It was planned for three Horsa gliders to land between each bridge, their six platoons to take out enemy guards, remove demolition charges and hold off counter-attacks for an estimated two hours until reinforced by 7th Bn., who at 0050 hours were to jump a mile east of the river. Spotting the bridges 3,000 feet below, the leading Horsa made for them descending on half-flap until, at a thousand feet, an arrester-parachute was released and the glider touched down less than 100 yards short of the bridge, breaking through the perimeter wire of the German defences. The other two gliders followed in and landed a few yards away.

Major Howard led the assault party of the Oxford and Bucks Light Infantry in a race for the bridge under a hail of Schmeisser machine-gun fire; one platoon over-ran the defenders on the far side of the bridge while the others cleared a pillbox and a network of trenches on the near bank. Of the three gliders heading for the Orne bridge, two landed within a few hundred yards of their objective which they secured without opposition. The demolition charges on both bridges were swiftly removed and linked bridgeheads formed astride the canal and river. The third glider touched down 15 miles to the east in the Dives valley.

At 0020 Hours, British Pathfinders dropped on D.Z.'N' north-east of Ranville; 'K' west of Troarn, and 'V' between the Merville battery and Varaville. Elsewhere sticks of pathfinders had not always been dropped accurately, although good enough to avoid the chaos that occurred with 82nd U.S. Division.

Dropped from C47s, 3rd Parachute Brigade were the victims of inaccurate delivery through failure to locate D.Z.s, or the inadequacy of Eureka pathfinder beacons. Several sticks of 1st Canadian and 9th Para. Bn. dropped near the River Dives intending to use the marshes as a natural flank barrier, but the Germans had flooded the area and many parachutists went straight into the water.

Landing before 0100 hours on D.Z. 'V' near Varaville, an area of orchards and fields divided by bocage, only about 50% of the widely scattered Canadians arrived at their rendezvous and moved off to capture and destroy the bridges over the river, before moving to the northern end of the Bois de Bavent.

Landing on D.Z. 'K' west of the Bures and Troarn bridges and well south of the other battalions, Pearson's 8th Bn., only 180 strong, utilised Engineer's jeeps to drive through the confused enemy to reach and destroy the two Dives bridges; then occupied the southern part of the Bois de Bavent. Here Colonel Pearson, a courageous and militant commander, used his battalion in an offensive patrolling role as though an outpost of a formidable defence that deterred the enemy from attacking.

Otway's 9th Bn. of Hill's 3rd Brigade had carefully rehearsed silencing the coastal defence guns at Merville; the attack being repeatedly simulated on a mock-up of the battery and its extensive defence works – just as Koch practised the assault on Fort Eben Emael. The battery, garrisoned by 130 men, was sited amid an elaborate defence system.

Three platoons in three Horsa gliders were to deal with the Germans in the actual gun positions, while the 9th Bn., dropping one and a half miles south-east of the objective, were in the roles of reconnaissance, breaching, assault and reserve parties. At 0420 hours when the breaching company were expected to be in position guided by mortar flares fired by the parachutists, the gliders were to descend in the neighbourhood of the battery. One of the three gliders broke loose from its tug before crossing the English coast, the second was unable to detect any landmarks or beacons and landed some three miles away. However, the third flew over the target and, releasing the arrester-parachute, crash-landed in an orchard only 200 yards from the battery site. The glider caught fire but its occupants escaped and dealt with the German patrol before taking up positions to hold off German interference with Otway's assault.

Before this, R.A.F. Lancasters bombed the battery with little effect, apart from devastating the village of Merville and killing most of the airborne reconnaissance party, but not before they had cut a gap through the outer wire and laid guide-tapes.

Dropping at 0050 hours over a large area, the 9th Bn. did not move off from their rendezvous until 0245 hours, and then with only 150 of the 700 men dropped, and lacking mortars, vehicles, artillery, engineer or medical personnel, mine detectors, and all the special assault stores; all that could be found was a single medium machine-gun, a few light anti-tank weapons and some radios. Using Bangalore torpedoes, the main party blew two lanes through the perimeter mine-field and, in the face of heavy fire passed through with sub-machine guns blazing to over-run the defences, when twenty German survivors surrendered. The four guns were immobilized with Gammon bombs, and yellow 'success' flares were lit. The fact that the guns were of smaller calibre than estimated should not detract from the quality of the 9th Bn.'s achievement, at a cost of 70 men. Re-assembling, the battalion took over a zone on high ground at Le Plein, later launching an attack an Sallanelles.

At 0300 hours, 47 Horsas and two Hamilcars appeared over Ranville, having lost 15 through A.A. fire and mechanical faults.

Two strips 1,000 by 60 yards and a Hamilcar strip 1,000 yards by 90 yards were cleared of obstacles and mines by sappers and marked by T-shaped beacons. They disgorged 7th Bn. reinforcements who moved to join up with Pine-Coffin's hard-pressed force at the Benouville and Ranville bridges.

At 2100 hours, on ground S.E. of Ranville cleared by 5th Bde. Engineers, Kindersley's 6th Air Landing Bde. came in in 246 gliders, including huge Hamilcars in action for the first time, transporting an armoured reconnaissance regiment equipped with Tetrarch tanks, Bren carriers and motor-cycles; the guns of 53rd Air Landing Light Regiment Royal Artillery, and 17 pdr anti-tank guns to strengthen the defences. The new arrivals were to take over close-defence of the bridgehead, allowing Generel Gale to bring 5th Parachute Brigade into reserve. In the days that followed, 6th Airborne Division fought hard and courageously to hold and extend their flanking position.

It can be claimed that the Division's performance on D-Day is the best-ever full-scale night airborne assault, and that the R.A.F. did exceptionally well in accurately delivering large forces in the face of anti-aircraft fire, difficult weather conditions, and without many of the expected navigational aids.

In the months that followed the Allied landings on D-Day no less than 16 airborne operations were considered too dangerous to commit airborne forces before the deadlock in the ground battle had been broken. But once that that had occurred, Allied armour moved so quickly that objectives were overrun before airborne operations could be launched or were required.

Operation Anvil, Southern France, 15 August 1944
(Also known as Dragoon).

On 15 August 1944, an amphibious invasion of Southern France was synchronised with the breakout of the main forces from the Normandy bridgehead. With the object of delaying enemy reinforcements reaching the coast, the 1st Airborne Task Force was formed under U.S. General Frederick, composed of five American Parachute Battalions, one American glider-borne regiment and the British 2nd Independent Parachute Brigade Group. Totalling 9,732 men, the force were carried in 535 C47s and 465

British and American gliders.

Despite accurate marking by early-dropped pathfinders, morning cloud and thick mist prevented concentrated delivery and men were scattered over twenty miles of countryside, fields and wooded hills in the Draguignan area. Only light opposition was encountered and no determined counter-attacks materialised. Perhaps the most interesting airborne aspect, and the only example of a direct parachute assault overwhelming beach defences, occurred by mistake when 29 plane-loads of infantry and artillery from 509 U.S. Parachute Infantry Regiment were dropped near German coast defences about three miles from St Tropez. Weathering an extensive naval and air bombardment as a preliminary to amphibious assault, the paratroopers got five of their guns into action and captured a German A.A.battery, two coastal batteries and 240 soldiers, then entered St Tropez ahead of the seaborne troops!

Operation Varsity, the Rhine Crossing, 24 March 1945

With Brereton's 1st Allied Airborne Army landing not *before* but *after* land operations had begun with Montgomery's 21st Army Group crossing the Rhine to capture Wesel, this was a new concept of employment for airborne troops. Cautiously respectful of the remaining German military capability, Montgomery was – 'definitely of the belief that the Rhine crossing could not be made without airborne support'. The Germans also appeared to believe that the river crossing would be staged in conjunction with an Allied airborne landing, subsequently massing in the area an estimated 712 light and 114 heavy anti-aircraft guns.

Ridgway's 17th Airborne Corps were to seize and hold the high ground between the rivers Rhine and Issel, north of Wesel; to defend the bridgehead against enemy counter-attacks and to prevent the enemy from sealing it off before enough troops were concentrated in it to continue the advance, and to prepare for offensive action to the east, under orders of 2nd Army. With 6th Airborne Division on the left (north) and 17th U.S. Division on the right, Ridgway planned to make a concentrated, direct attack against objective areas, with the entire assault force delivered in one lift as quickly as was practical – parachute troops to be landed within twenty minutes from two parallel divisional streams, followed by the gliders of 6th Division within 45 minutes and those of 17th Division within 108 minutes. By flying in tight formations nine planes abreast, U.S. IX Troop Carrier Command could drop a parachute battalion in two minutes and a brigade or regiment in ten.

The drop was scheduled for 1000 hours because – (a) The anti-aircraft fire threat was greater on the Rhine crossing than in any previous airborne operation, and the powerful German night-fighter defences presented tactical as well as navigational objections to a night descent. (b) The airborne landings were within range of day-light support by 2nd Army medium and heavy artillery.

Soon after daybreak on 24 March 1945, the British and American airborne divisions took off from England for their rendezvous near Brussels, from whence they were to continue in parallel columns. The airlift was composed of 1,696 transport planes and 1,348 gliders carrying 21,680 Allied airborne troops in a single lift, escorted by nearly 1,000 fighter aircraft with twice that number supporting the ground operations. Six hundred of the gliders were on double-tow. The fly-in lasted two hours and 40 minutes, with three columns, each nine aircraft or double-tow gliders across, moved in an air armada over the river. They were met by devastating streams of anti-aircraft fire that destroyed 44 transport planes and 80 gliders, with 440 troop-carrier aircraft badly damaged.

Flying separately from the main American force because of their superior speed, were 72 C46 'Commando' aircraft, a new and larger troop carrier lifting 30 troopers, with Col. J. Coutt's 513th Parachute Regiment, jumping from doors on both sides of the fuselage. Twenty-four of these planes went down in flames, from all but one the paratroopers got out. It transpired that '. . . the C46 was a fire-trap – if a wing tank was punctured the fuel ran down along the fuselage, and an incendiary shell could set the aircraft aflame in a second'.

Despite haze and smoke, unit-drops went as planned, the 513th being the only inaccurately delivered formation. Landing on one of 6th Air Landing Brigade LZs, they immediately went into action alongside British glider-troops, taking on and neutralising 88 mm guns in the area, before using their fire-power to break out southwards towards their allotted objective.

The concentration and speed of delivery achieved by tight formation flying meant that each group of aircraft was only overhead for a short time, so reducing the effectiveness of anti-aircraft fire; also it enabled the airborne invaders to overwhelm ground resistance with minimised casualties even after fierce fighting. Even so, with many DZs and LZs in clearings of the wooded area men hung from trees by their harnesses, presenting easy targets; losses were incurred on the DZ's from light A.A.weapons used as field-guns. However, no units took casualties on the same scale as German battalions in Crete.

Poett's 5th Parachute Brigade dropped from about a thousand feet, taking casualties from airbursts and small arms fire on the way down. The 7th, 12th and 13th Bns. landed astride the road leading into Hamminkeln and by late afternoon captured their objectives. The gliders of 6th Air Landing Brigade came in through a haze of smoke and dust raised by the artillery bombardment, taking 27% casualties, their passengers suffering losses on a similar scale. The gliders carrying 17th U.S. Airborne Division, being towed at only 600 feet, were in free flight for a much shorter period and their losses were fewer, although more tug-aircraft were shot down.

Troop-carrier and tug aircraft frequently turned for home with flames and smoke pouring from them – General Gavin recorded seeing 23 planes burning in sight at one time. The crews realised they were too low to use parachutes and stayed with their stricken craft or, forgetting to remove anti-flak-waistcoats, plunged to their death through parachutes not operating beneath the heavy armour.

Despite fierce defensive fire and desperate resistance on the ground, the great accuracy of the assault were quite overwhelming and all objectives were taken and link-ups with the leading elements of 2nd Army were complete by 1500 hours. Nevertheless, operation 'Varsity' was not entirely necessary and, although the Allied airborne troops accomplished their task, the success did not justify the losses of 41 air crew killed, 153 wounded and

163 missing; 6th Airborne Division lost 347 men killed and 731 wounded; 17th Airborne Corps had 159 killed and 522 wounded.

GERMAN AIRBORNE OPERATIONS DURING THE ARDENNES OFFENSIVE, DECEMBER 1944

General Student organised two battalion groups for airborne operations in advance of the armoured thrust in the Ardennes. Parachute-trained men were taken from units in 1st Parachute Army to make up battalions under Obersts von der Heydt and Herrmann. There were 67 Ju 52's available, but operations in Russia had taken most of the experienced troop-carrier pilots, and the majority of the newly assembled crews were insufficiently trained in night navigation with no experience of dropping paratroops. Heydte's group were to drop on Mont Rigi, near the Malmedy-Eupen-Verviers crossroads, to prevent the movement of Allied reinforcements; they were to remain in control until relieved by Dietrich's 6th Panzer Army. Delayed 24 hours, this final German airborne operation was ill-conceived and badly executed so that the only *night* drop of the war became a failure. Poor weather conditions and the pilot's lack of experience resulted in only a third of the aircraft reaching the correct dropping-zone, where a 30 knot wind caused many injuries among the paratroopers. Many of the badly injured, unless fortunate enough to be captured by the Americans, lay on the snow-swept hillside to die of wounds or exposure. Streaming south, American troops immediately attacked the survivors of the airborne force, who having to fight as soon as landing, were unable to establish a blocking position. Von der Heydte, jumping with a splinted arm from a previous injury, was left with only 200 men to harass the roads. In an attempted breakout to meet the advancing German Panzer force, the majority, including Von der Heydte, were captured, and the battalion was virtually destroyed. Herrmann's group, briefed to jump and establish a bridgehead over the Meuse, had their operation cancelled when the Ardennes offensive lost its momentum.

The French airborne activities, April 1945

Towards the end of World War Two, the French 2nd and 3rd

Regiments de Chasseurs Parachutistes were trained for offensive operations in small parties who, by avoiding German counter-measures, did not have to rely on concentration to carry out their tasks. This was in contrast to earlier raids in company of battalion strength which, despite accurate concentrated drops, ended in failure because the enemy overcame their initial disadvantage of being surprised and concentrated superior forces to defeat the raiders. A classical example of the new technique occurred in April 1945 when the two Regiments, totalling 700 men including a Special Force liaison team, dropped ahead of the advancing 1st Canadian Army in north-east Holland. Their task was to disrupt German defences; rally Dutch resistance; guide and inform the advancing Canadians; remove charges and prevent the demolition of 18 bridges, and to secure Steenwijk airfield for R.A.F. use. A link-up with ground troops was expected within 72 hours.

Drops were made onto 19 dropping zones without Pathfinders or reception parties, blanketing clouds forcing 'blind' drops from 1,500 feet, using fixes from Bomber Command 'Gee' navigational system that enabled air crews to accurately pin-point their location over enemy territory. All but one of the 47 Stirling bombers successfully dropped their loads; the remaining aircraft dropped its passengers on the following night. Six parties landed with complete accuracy, the remainder were within two to four miles of the target – satisfactory for this type of operation of self-contained 15-men parties not needing to establish contact with other groups.

THE GERMAN INVASION OF CRETE

The only true strategic operation involving parachute forces attempted in World War II, Crete was the first *land* battle won by airborne power alone and the first *naval* battle won entirely by air power – by attacking an island from the air before winning command of the surrounding seas, German General Student demonstrated the potential of airborne warfare in a manner that has never been equalled.

Averse to leaving Crete in Allied hands because of the threat posed to the Rumanian oilfields at Ploesti, Hitler lacked both time and facilities to mount an amphibious operation in the face of Royal Navy command of the seas. Subsequently Student convinced his leader that Crete could be taken from the air, and Operaton 'Merkur' (Mercury) was planned for mid-May.

From west to east 160 miles long and about 20 miles wide, Crete is marked by four chains of steep and barren mountains most of which descend almost straight down into the sea, and its rivers flow through deep valleys forming obstacles to lateral cross-country movement. In the north, where the land slopes more gradually to the sea, there are long narrow coastal plains with the island's best ports at Suda Bay, Retimo and Heraklion. At Heraklion, 70 miles to the east, was the main airport, watched over by twin conical hills known to British troops as the 'Charlies'. Close to the sea and dominated by inland hills were two more military airstrips at Retimo, between Heraklion and Canea,

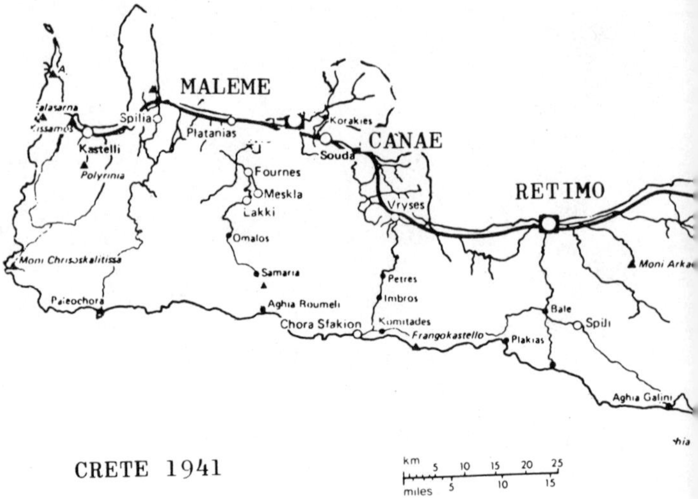

CRETE 1941

and at Maleme ten miles west of Canea. Roads were very bad and air reconnaissance was difficult because of the thick cultivation of the northern plain.

In mid-May, the island's garrison consisted of a 5,000 strong British Infantry Brigade (2nd Black Watch; 2nd Yorks and Lancs, and the Leicesters); a Battalion of the Welch Regiment, the Northumberland Hussars with twenty-two old and unreliable tanks, a troop of artillery and some Royal Marines, plus some lightly armed and poorly trained Greek and Cretan units. But the German 7th Air Division's Intelligence Summary No. 4 dated 19th May 1941 (the day before the invasion) was wildly inaccurate in that it completely omitted some 25,000 men from the 6th Australian and 2nd New Zealand Divisions evacuated to the island from Greece. Although only armed with their personal weapons, they were a strong reinforcement and raised the garrison strength to 40,000 men. However, it was a force particularly short of mortars, medium machine-guns, tanks and field artillery; its transport was quite inadequate and there was an almost total lack of radios making communication between formations

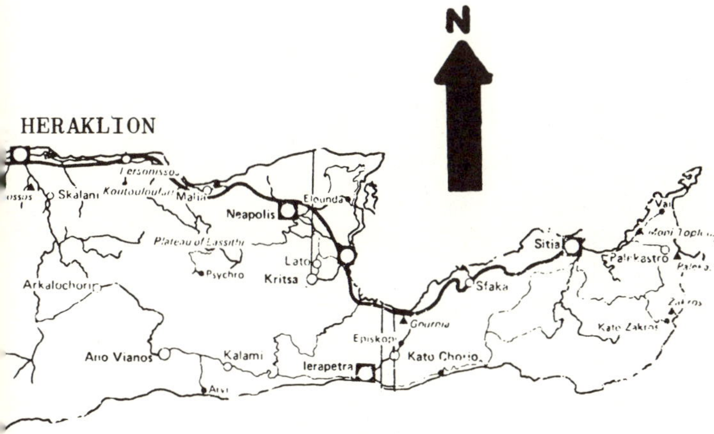

and headquarters almost non-existent. Major-General Bernard Freyberg V.C. who came to Crete from Greece with the New Zealand Division, was in command.

The Luftwaffe ruled the air and the three airfields were exposed so as to preclude dispersion or defence in depth; the few R.A.F. aircraft were so heavily out-numbered that they were eventually withdrawn. Allowed freedom of action by this air superiority, the Germans were able to prevent the movement of defending troops during daylight. This meant that relatively small airborne forces could hold objectives, as they had done in Holland in 1940.

Freyberg had too little time to remedy faults and deficiencies in his troops and their equipment, or in their deployment and communications. Believing that an island surrounded by sea must be attacked amphibiously, the New Zealand commander deployed two-thirds of his troops in a coastal-defence role and the remainder to cope with airborne attack. He did not fully appreciate that the three airfields were the only vital points on the island. Nevertheless those troops that were deployed around them were a real threat to German plans and the New Zealand commander's insistence on concealment, night vehicle movement and well dug defensive slit-trenches, preserved the defenders from heavy casualties during preliminary attacks, whilst deceiving the enemy into thinking the island thinly held. In fact, for the first eight hours the overall ratio was to be 4,300 attackers against 40,000 defenders, although on the objectives the odds were less. The material weakness of the defence – broken communications, immobility, low fire-power and little armour, caused the defenders to be lightly equipped infantry fighting highly trained and well-organised elite troops. Nevertheless, if Freyburg, local commanders and their troops had seized their opportunity a painful defeat could have been inflicted upon the Germans and Crete held.

General Student had at his disposal the XI Air Corps, consisting of Sussman's 7th Air Division of three parachute regiments, each of three battalions; General Meindl's Assault Regiment of four battalions; General Ringel's 5th Mountain Division from Italy (taking the place of the 22nd Division who were in Rumania) and reinforced to four regiments each of three battalions; and a Panzer battalion. Corps troops included artillery, anti-tank, engineer, signal and supply battalions and an airborne field hospital. The German 7th Air Division and the Assault Regiment consisted of some 8,000 paratroopers, and 5th Mountain Division 14,000 men who could not take part in operations until landed on a captured airfield.

Lacking time to provide tropical dress, these troops fought throughout in heavy uniforms, despite intense heat.

The invasion force were transported by General Conrad's XI Air Command of ten troop-carrier groups each of 50 Ju52s, and 80 DFS230 gliders, supported by 700 fighters, bombers and reconnaissance aircraft of General Richthofen's VIII Air Corps. The Ju52 carried 18 men in the air-landing role, but only 13 men were lifted in operational parachuting, to allow for weapons containers and space for the paratroopers to move to the door and evacuate correctly. So the 500 Ju52s could lift 6,500 paratroopers, reduced to about 5,000, because some were only dropping supply containers. This meant that the 8,100 airborne troops had to be carried in two lifts, the second dropping eight hours after the first.

The Operation Order for Mercury read 'XI Air Corps supported by VIII Corps will capture the Island of Crete and hold until relieved by Army Troops (German 5th Mountain Division)'. Student planned to attack all three airfields so as to be reasonably sure of capturing at least one in serviceable condition for the air-landing of his 5th Division. In the early morning the first of the two lifts were to set down the Western Group to seize Maleme airfield, and half of the Centre Group to threaten Canea; the second lift in the afternoon would drop the other half of the Centre Group onto Retimo, and a third (Eastern) Group against Heraklion. Western Group was basically the Assault Regiment under Meindl with its 1st Battalion glider-borne and two of its companies detached outside the Group for other missions. Meindl planned surprise landings by his two glider-company groups, commanded by the redoubtable Major Koch, followed fifteen minutes later by three parachute battalions dropping near to but not on the objective.

The Centre Group, commanded by Sussmann, consisted of Heidrich's 3rd Parachute Regiment, two glider companies from the Assault Regiment, Divisional Engineer and anti-aircraft battalions; and later 100th Mountain Regiment, part of whom were to come by sea.

The remainder of the Centre Group was to land at Retimo with the second lift – Sturm's force of 1st and 3rd Bns. of the 2nd Regiment plus ancillary detachments. The attack on the airstrip (close to the coast six miles east of Retimo town) was to be by a drop directly on the objective, with Major Kroh landing immediately east of the field with 800 men of the 1st Bn. and capturing it, while a 800-strong group from the 3rd Bn. dropped halfway between the air-strip and the town, which they were to attack from the east. Regimental Headquarters were to land between the two groups, close to the sea and just west of the strip, and control the two battles from there.

Eastern Group under Ringel consisted of 1st Parachute Regiment, 2nd Bn. of 2nd Regiment and 5th Mountain Division (less 100th Regiment) but only the paratroopers under Brauer were to land in the assault phase, the follow-up force to be air-landed as soon as Heraklion airport was captured. The 2nd Bn. were to land west and south-east of the field and take it by direct pincer attack and the 3rd Bn. split between dropping zones west and south of the objective. The remaining battalions were to be used on the flanks – 1st Bn. three miles east of the airfield, 2nd Bn. from 2nd Regiment two miles west of Heraklion. As soon as the airfield had been taken, Mountain troops were to fly in and mop up, supported by heavy weapons landed from the sea.

Providing the entire German XI Corps could be landed, their air superiority would probably give them the victory, as the defenders on an island were unlikely to receive reinforcements. On the other hand, only half XI Corps could land before an airfield was secured, and only half the assault force were in the first wave. The Luftwaffe could turn the balance by providing them with air artillery, adequate reconnaissance and plentiful supplies during daylight hours. Operation Mercury also included a mixed flotilla of 25 to 30 ships each of 400 tons, with some larger vessels capable of carrying a single tank. The crews were Greek, as naval manpower resources allowed only one German sailor per ship. The Italian Navy provided two torpedo boats as escorts. This force was incapable of mounting an opposed amphibious landing but provided the means of transporting light armour, artillery, supplies, heavy equipment and their crews, etc., onto an already secured port or beach.

The German airborne force that took off early on 20 May 1941 consisted of the first assault parties in 53 gliders towed by Junkers 52s; in five gliders, the Headquarter Staff of 7th Air Division flew behind the mass of paratroop aircraft, but seemingly loaded beyond safety limits, one broke up and crashed, among the dead was General Sussman.

The clear and bright dawn saw the defenders of Crete coming under air attack, as they stood-to, particularly at Maleme where it was severe. This preliminary German air bombing, by almost completely silencing the air-defence batteries, prevented the airborne attackers from heavy casualties and only seven troop carriers were lost out of 493 on the first lift.

At Maleme Brigadier Puttick, the New Zealand Divisional

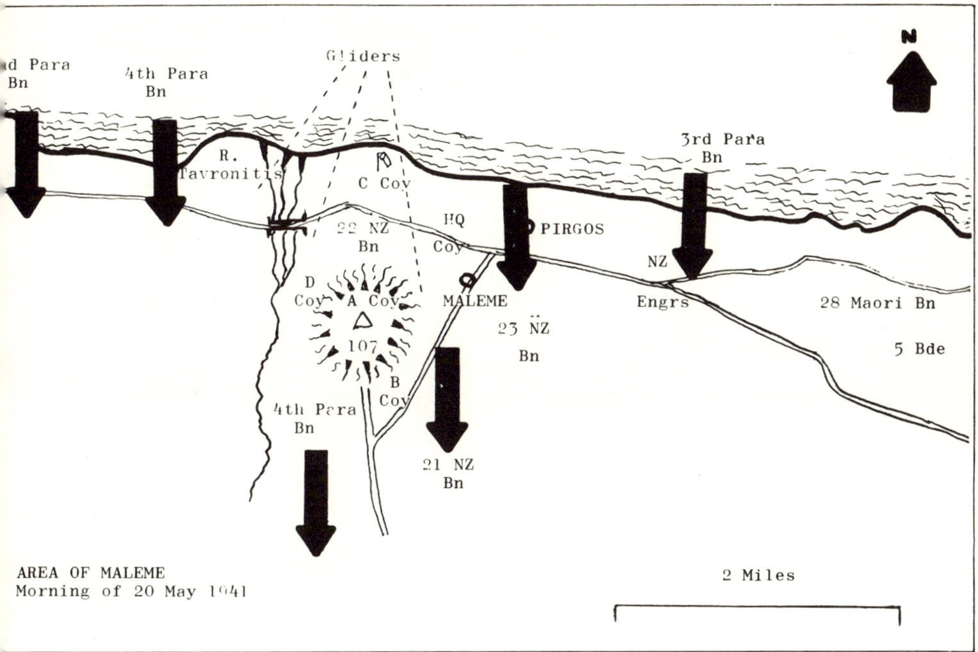

Commander, allocated the 22nd N.Z. Inf. Bn. to the sector east of the Tavronitis River, spread widely around the airfield, with a company on Hill 107; one on its western lower slopes; one on the airstrip, and a fourth in depth south-east of Hill 107; Headquarters Company was in the village of Purgos. The rest of 5th N.Z. Brigade, 21st and 23rd N.Z. Bns. and 28th (Maori) Bn. with an Engineer detachment, were concentrated nearby to the east.

In this area, on a rocky, broken and tree-covered terrain quite unsuitable for glider landings, Nos. 3 and 4 Companies of Koch's 1st Bn. Assault Regiment landed; gliders coming down near the beach, at the mouth of the Tavronitis River, further inland beside the bridge, and on the south-western eastern slopes of Hill 107. The terrain caused several to crash, others suffered from heavy machine-gun fire as they came in or landed. When their occupants poured out they were met by a hail of accurate fire, mainly from the summit of Hill 107, that forced them to take cover and begin a fire-fight with the defenders. Major Koch was wounded and his force were soon reduced to about 100 men, less than half their original strength, hanging on grimly at the extreme western end of the airfield in the area of the Tavronitis River. Major

The Crete Wargame. Rifleman and light machine-gunners of Ringel's 5th Mountain Division fighting in the area of Maleme airfield.

Scherber's 3rd Bn., planned to drop on or close to the beach, fell further south among the New Zealand positions, and lost 400 out of 600 men within 45 minutes, including the Battalion commander.

West and south of Maleme airfield, the 2nd and 4th Bns. (less two detachments) were able to land without being strongly opposed. Meindl's R.H.Q., landing just west of the Tavronitis, immediately sent two companies from the 2nd to make a wide outflanking movement and come up upon Hill 107 from the south. The 4th Bn., a company of the 2nd and the remains of Koch's force, some 700 men, consolidated their positions on the river line on the western edge of the airfield. With Meindl severely wounded, and the defenders stubbornly resisting, coupled with heat, thirst and difficulties in distinguishing friends from foe in the dust haze, it seemed that the Germans' direct attack on Maleme had failed.

The remainder of Lift One (half of the Centre group) were also in action at Canea and Suda Bay, where the gliders landed ahead of the paratroops, who were delayed by navigational errors and anti-aircraft fire so that they arrived scattered and with some sections isolated. Altmann's took heavy losses from the Welch

The Crete Wargame. A mountain gun, air-landed on the temporary strip at Heraklion, in action under fire.

The Crete Wargame. German paratroopers of Derpa's 2nd Battalion attack the outskirts of Galatas.

Regiment and the Northumberland Hussars, and were eventually completely destroyed. Attacking the heavy guns south of Canea, Genz's No. 1 Company overwhelmed the 180 gunners (who were without personal arms) before being pinned down by Royal Marine detachments.

While the glider-borne troops were fighting grimly, the 3rd Parachute Regiment approached its drop-zones. Von de Heydte's 1st Bn. landed intact on the southern end of their Regiment D.Z., moved forward through the undefended prison buildings* eastwards towards Perivolia, where they encountered opposition and were held up. Derpa's 2nd Bn. landed south-west of Galatas and took about 150 casualties in the initial fighting, but succeeded in threatening the New Zealand positions on the Galatas Heights that covered both Canea and the road linking the town with Maleme.

*After the war, Von der Heydte wrote – 'It is possible to land among groups of houses . . . on roofs. The paratrooper must be able to cling to the roof with the aid of grappling hooks and quickly cut an opening in the roof so that he can make his way into the house'.

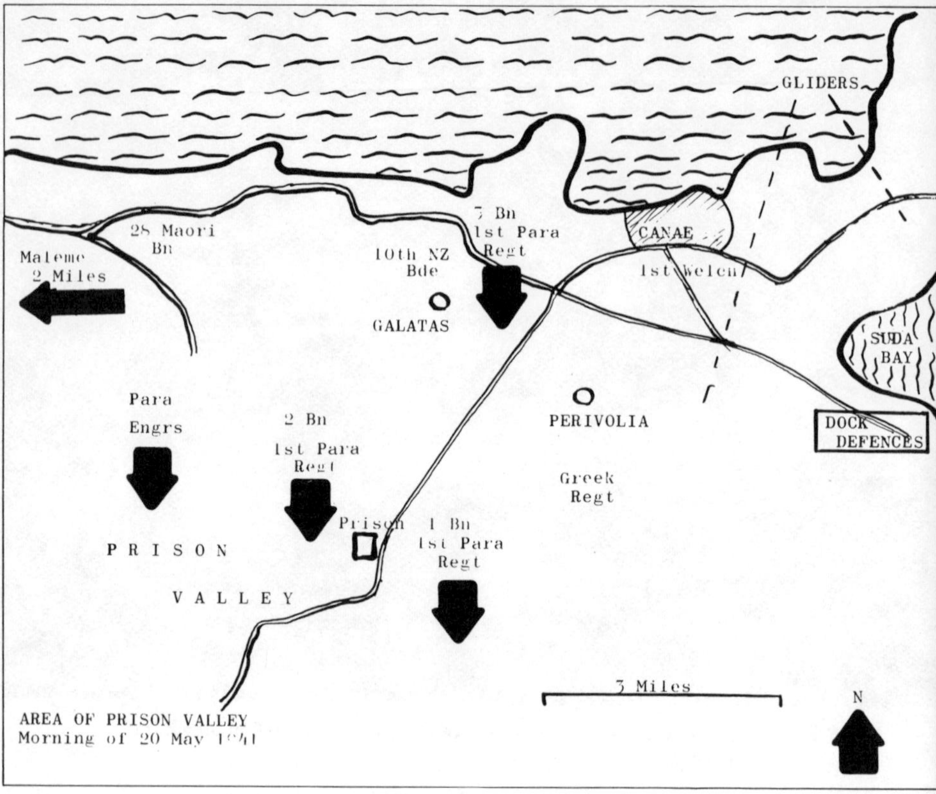

AREA OF PRISON VALLEY
Morning of 20 May 1941

Heilmann's 3rd Bn. using a plane-load type delivery, arrived scattered both by time and distance and, being attacked piecemeal by the defenders, were unable to achieve their objectives of the Canea—Maleme road and the Galatas. By nightfall, this battalion had ceased to exist as a unit. Landing west of the prison, the engineer Battalion suffered more than 100 casualties when they came up against Greek and Cretan troops.

Late in the day two companies of the 2nd Bn. had gained a foothold on high ground near Galatas, but Heidrich (who had taken over command of Centre Group) realised he was virtually confined to Prison Valley. Improvised airfields in Greece and Greek Resistance interference with communications delayed Lift Two, so that the preliminary air-bombardment was over well before the lift came in, and the troop-carriers were hotly received by anti-aircraft and machineguns from defenders who had got over the shock of the air bombing, and were aided by there being no smoke and dust clouds to conceal the landings. Dug-in on and around Heraklion airfield, the Black Watch concentrated their fire on each wave in turn, having time to take aimed shots at almost every individual descending paratrooper. Dunz's 2nd Bn. 1st Regiment, landing west and south-east of the airfield, suffered more than 400 casualties within an hour of landing, completely failing to achieve their objectives. Expecting to meet only infantry, the Germans were surprised to come up against half-a-dozen tanks and sixteen Bren carriers; many paratroopers were killed by the Black Watch before they could find their weapon-containers as the British troops, realising the importance of these cylinders, prevented the attackers from reaching them.

At this time the Germans had yet to develop techniques of dropping rifles and machine-guns with individual soldiers, but sent them down in cylindrical containers to be collected on the dropping zone. Men jumped with Schmeisser sub-machine guns and Luger pistols both firing 9 mm pistol ammunition with limited range, accuracy and stopping power. At Crete German airborne troops suffered gravely by depending upon containers for the delivery of weapons.

The 3rd Bn., 1st Regiment did not take such heavy landing casualties and, although facing fierce opposition from Greek troops and a company of the York and Lancaster Regiment, succeeded in breaking into Heraklion town and by nightfall were threatening the quayside. Two flanking battalions landed in undefended areas – 2nd Bn., 2nd Regiment (less two companies left behind through a breakdown in Lift Two arrangements) blocked the main coast road to the west; to the east, Brauer and R.H.Q. landed safely with 1st Bn., dropping over a three hour period so that it was midnight before any of them reached the battle area. At midnight the Germans were split – one battalion on each eastern and western flank of the British sector.

At this time the Black Watch securely held the airfield; the area between the field and Heraklion itself was heavily defended by 2nd/4th Australians, 2nd Leicesters, 2nd York and Lancasters and other British troops, while Greek and British troops were grimly pressing to eject paratroopers from within Heraklion.

The Crete Wargame. Buildings on the outskirts of the village of Galatas in the Prison Valley area being defended by 10th New Zealand Brigade against the attack of Derpa's 2nd Parachute Battalion.

The Crete Wargame. Men of the 14th New Zealand Infantry Brigade defend the village of Galatas in the Prison Valley area.

Sturm's Centre Group of 1,380 men formed the second lift with Retimo as their objective. Here, the defenders of the high ground inland from the coast road completely dominated the airfield, and two experienced Australian battalions – the 2nd/1st and the 2nd/11th, were skilfully concealed on these overlooking hills midway to Retimo, which was held by armed Cretan police. Two poorly armed and under-strength Greek Battalions filled the gap between the Australian units.

Dropping slowly and dispersed, the German 1st Bn. was scattered over a five mile area; two companies were immediately engaged by the Australians and took heavy casualties, one losing all its officers. Dropping inaccurately, the 3rd Bn. landed virtually unopposed east of Retimo and about 400 of them attacked the town in the face of stiff resistance from the Police. The German Headquarters Group descended at the feet of the Australian left flank position, within range and vision of almost every weapon that could be brought to bear by the 2nd/11th Battalion. The survivors of this drop quickly took cover and formed into groups, but were dispersed and many taken prisoners as the Australians came down and attacked them.

After an hour Sturm's force was reduced to about 1,000 men, with the majority of them in the 3rd Bn. taking the least important objective at Retimo; others were hiding in small groups beneath the Australian positions. Major Kroh collected those paratroopers who had dropped wide and joined the two 1st Bn. companies battling fiercely with the Australians on the vital hill overlooking the airfield. The Germans made a skilful attack and, in spite of the defenders being determined and well positioned, killed or drove them onto the southern neck of the hill; a quickly mounted counter-attack failed. Despite its poor start, the 2nd Regiment's attack on the airfield was beginning to succeed, and had the 3rd Bn. left Retimo town and attacked from the west, the heights overlooking the airfield, the Germans would have gained their objective.

His communications working reasonably well (although a Fiesler 'Storch' liaison aircraft sent to bring back information from Retimo was immediately captured by the Australians as it landed) Student decided that his best and perhaps only hope of seizing an airfield lay at Maleme. So he decided to switch his main effort to secure that airfield and fly in the 5th Mountain Division. If an airfield was not soon taken the entire 7th Airborne Division could be lost; even if the seaborne invasion reached Crete safely they would come up against the same determined opposition that had frustrated the paratroopers.

Student had planned to commit almost all his paratroopers on the first day, but part of his anti-tank battalion had been held back and constituted a parachute reserve. Together with troops delayed by normal mechanical failures in both air lifts and a breakdown in the arrangements of the second lift, this formed a reasonable reserve, including three complete rifle companies available in Greece to be sent to Crete as reinforcements.

At this stage, the defenders had accomplished the first two of the three essential defensive duties against airborne attack –

(1) the maximum casualties had been inflicted at the moment of landing.

(2) the attackers' objectives had been ascertained and held.

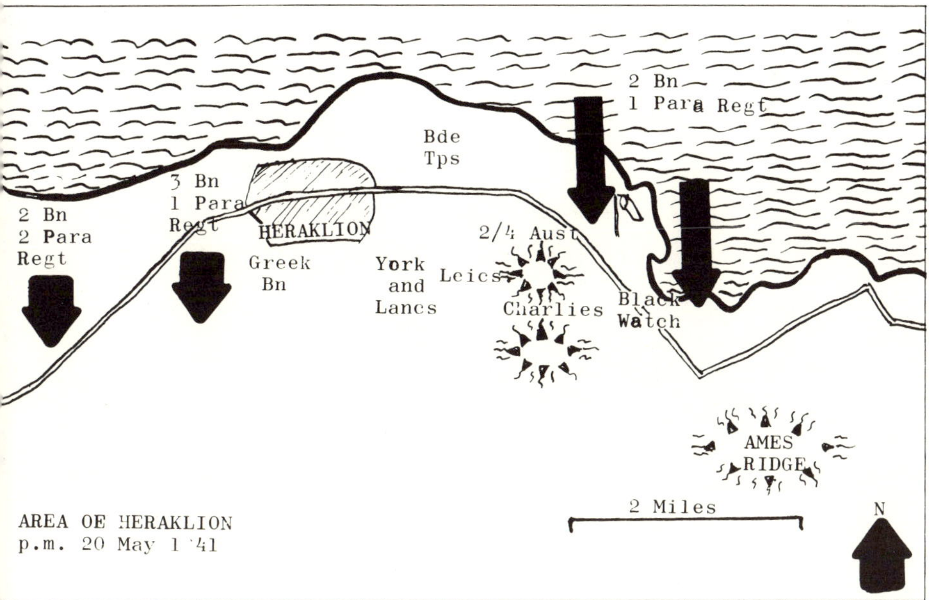

AREA OF HERAKLION
p.m. 20 May 1941

(3) the eventual destruction of the airborne invaders.

If the defenders could retain their grip and, by taking the offenders, destroy the invaders, then they would win. In fact, the German invasion could have been beaten at the very outset had the defenders immediately and energetically attacked the paratroops as they landed, rather than limiting themselves to purely defensive measures.

In the Heraklion area, although Brauer's group failed to achieve their mission it succeeded in containing a strong Allied force which could usefully have reinforced the Australians at Retimo or the New Zealanders further west. Despite Brigadier Chappell's 14th Brigade being some 3,000 strong, including tanks and light supporting weapons, against about a thousand German airborne troops split into two groups, it was the latter whose aggressive demonstrations pinned down the superior Allied force. At one stage, Brauer sent an urgent message for mountain artillery which could not be dropped and had to be air-landed. As

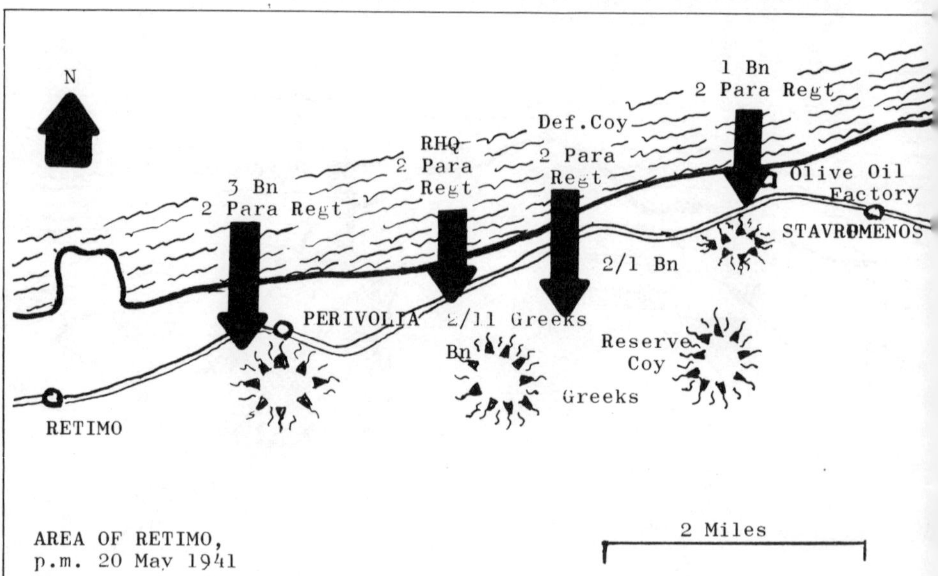

AREA OF RETIMO,
p.m. 20 May 1941

The Crete Wargame. German glider-borne assault troops and paratroopers land under fire from entrenched troops, supported by a light tank of unknown vintage and carriers. See also photograph on page 118.

the airfield at Heraklion was not in German hands, paratroopers fighting in the sector built a landing strip behind a hill and the artillery were safely landed.

At Retimo, Campbell's two Australian battalions displayed sound leadership, tactical planning and courage in overcoming the fierce resistance of the paratroopers as, assisted by Greeks, they cleared up the defensive sector, killing or capturing the majority of them including Sturm and his Staff. German 1st Bn. survivors held out at an olive-oil factory at Stavromenos, a mile and a half from the airfield, while remnants of 3rd Bn., well to the west, were kept on the defensive trapped between Cretan police and 2nd/11th Australians.

In the Prison Valley area the village of Galatas was the key objective; the 10th N.Z.Brigade firmly held on to their defences and they might well have prevented the threat of the Centre Group dividing Freyberg's and Puttick's attentions between Galatas and Maleme. At the latter place, the night 20/21 May found the New Zealanders firmly holding Hill 107 and most of their original positions; the two German companies approaching Hill 107 from the south had only gained the lower slopes. A New Zealand counter-attack on the bridge area by a platoon of infantry and two supporting tanks was repulsed with the loss of the tanks, but it worried the Germans who reasoned that one of the three New Zealand battalions that had destroyed Scherber without any loss to themselves could be directed against the Assault Regiment's foothold. But faulty communication and misunderstanding of the situation by the New Zealand commanders caused an unnecessary abandonment of the vital Hill 107, and at dawn, when the exhausted Germans expected to be grimly attacked, they found that the New Zealanders had pulled-out.

In this area, before the evacuation of Hill 107, the Germans had almost surrendered the initiative and had the defenders of Crete seized the opportunity to slam the two doors at Retimo and Maleme, the battle would have been won. It was done by Colonel Campbell's Australians at Retimo, but at Maleme the chance came and went unrecognised.

Student sent a Ju52 to land on a sandy stretch of the airstrip at its western end, invisible to defenders, carrying a staff-officer to assess landing and taking-off facilities. This reconnaissance revealed that air-landings in strength might be practicable at Maleme, so at 0800 hours six supply aircraft landed on the beach near the mouth of the Tavronitis two miles north-west of the airstrip. At the same time, two parachute battalions were dropped in the defenders rear near the coast just east of Pyrgos, but they landed directly on top of 5th Brigade units and only 80 succeeded in joining Gericke now pushing on steadily eastwards from Hill 107. Student ordered the Mountain Battalion to be air-landed regardless of aircraft losses; the first Junkers 52s came in at 1700 hours, bringing Oberst Ramcke to take command of the paratroopers at Maleme. On 21 May, the Germans had enlarged their foothold there and on succeeding days were quickly reinforced by Mountain troops being air-landed during daylight hours.

Freyberg, still preoccupied by the danger of amphibious landings, did not recognise the real danger, so that the two battalions he ordered to counter-attack west astride the coast road were too

The Crete Wargame. At Heraklion airfield, German assault troops landing from gliders and paratroopers meet opposition from the entrenched Black Watch around the airfield and face a counter-attack from light tanks, carriers and infantry.

The Crete Wargame. The Black Watch counter-attack at Heraklion airfield, supported by an old British Light tank and Bren carriers.

few and too late. Moving towards the airfield in the early morning on May 22, this force were so strongly resisted by the survivors of Scherber's 3rd Bn., decimated on the first morning, that by dawn the counter-attack was well short of its objective when broken-up by Luftwaffe attacks. It was at this moment that, perhaps ironically the Royal Navy were destroying or scattering the German amphibious force.

With Maleme as their main effort and the groups at Heraklion and Retimo merely containing local garrisons, throughout the 22 May the whole of 5th Mountain Division were landed on an airfield littered with burning and wrecked aircraft. Prisoners and captured tanks aided the unloading of supplies and clearing the wreckage. East of the Maleme strip, the full strength of the Luftwaffe completely immobilised the New Zealanders during daylight and silenced their artillery. The battle to open the door to Crete was over, the Germans now merely had to build up to full strength and destroy the Allies piecemeal in a systematic fashion as, in Student's words, 'They rolled up Crete from the West'.

Utz's 100th Mountain Regiment were sent south-east on a wide outflanking attack; Galatas was taken early on 26 May after twice changing hands and Heidrich, with the remnants of Ramcke's Assault Regiment and the fresh 5th Mountain Division, rolled steadily forward. They captured Canea on the 27th, forcing the Allied forces that had defended the Maleme—Canea—Suda Bay sectors to conduct a fighting withdrawal across the White Mountains to the tiny port of Spakia on the south coast.

On the 28th, a German force, including two tanks that had been landed by sea, proceeded along the now open road to pass Suda bay towards Retimo to attack Campbell. With no orders for withdrawal, the Australians were still fighting bravely but, short of ammunition and supplies, could not withstand this further attack and broke up into small groups who made for the south coast as the remainder surrendered. Sturm and 500 paratroopers were freed, but some 700 Fallschirmjager lay dead on the ground.

At Heraklion, Chappell's Brigade (less the Greeks for whom there was no room) were evacuated by the Royal Navy on the night of 28 May; 14th Brigade suffering badly when their ships were attacked by the Luftwaffe, before reaching Alexandria they

The Crete Wargame. German paratroopers land under fire from entrenched troops at Maleme and go into immediate action.

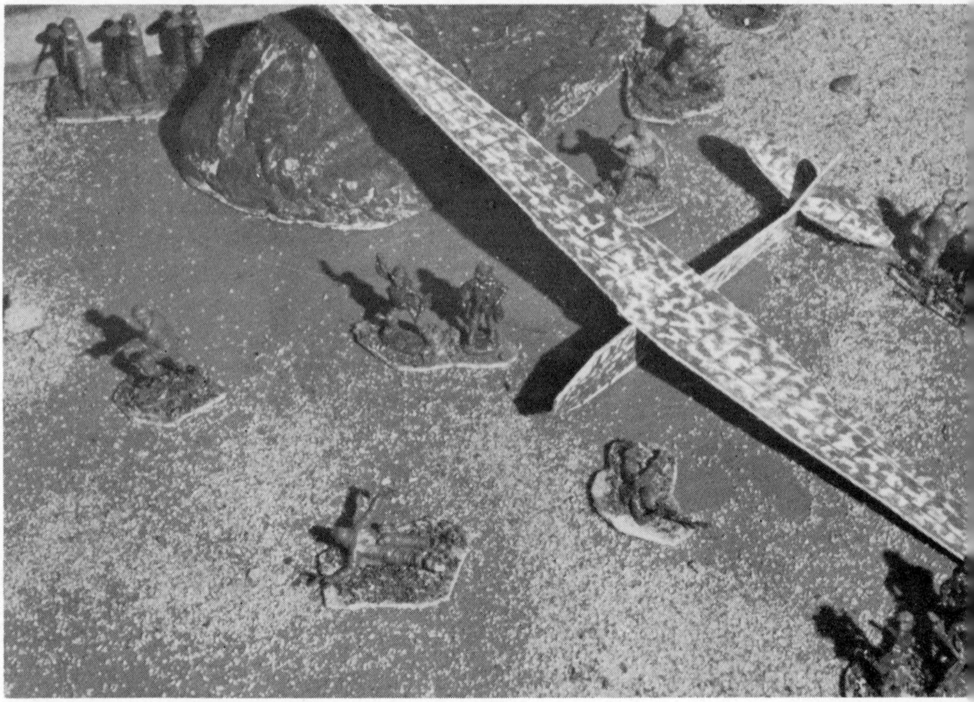

The Crete Wargame. A glider of the German centre group at Canae disgorging assault troops under fire.

lost 800 killed or wounded.

Commandos and infantry fought a skilful rearguard against relatively light German pursuing forces on the mountain roads to Spakia from where courageous and determined Royal Naval efforts evacuated as many as possible before heavy losses inflicted by the Luftwaffe forced the operation to be called off with 5,000 troops left ashore.

The German success on Crete was due to a high standard of leadership, courage and initiative at all levels, particularly among the paratroopers themselves who, coming up against resistance far tougher than expected, might easily have become discouraged and surrendered. Suffering severely during their drop, scattered and isolated survivors in seemingly hopeless positions could well have given themselves up but, despite their undoubted inner fears, they continued to fight bravely, their leaders and individual soldiers demonstrating that their training had equipped them to make sound tactical assessments and act positively and courageously.

At Crete, the Germans lost 4,000 killed and not less than 2,500 wounded; 170 Ju52s and about 40 combat aircraft were lost. Including naval casualties, the British lost 4,000 killed, 2,500 wounded and 11,800 prisoners; nine warships sunk, 17 damaged and 46 aircraft destroyed. 400 Palestinian and Cypriot Pioneers were captured and 10,000 Greek and Cretan armed forces or police became prisoners.

The losses suffered by the 7th Air Division and the Assault Regiment so horrified Hitler that he decreed: The day of the parachutist is over'. So Crete, a historical viewpoint and crowning glory of German airborne achievement, marked the end of a dream for the German airborne forces. Despite its brilliant conception and daring execution, the battle forced Student to lament that 'Crete was the grave of the German paratroops'. Thus, with achievements of German airborne troops at their highest level, the initiative in airborne warfare began to slip into Allied hands and, this German rejection gave an indirect victory to the courageous and stubborn Allied defenders of Crete.

ARNHEM — OPERATION 'MARKET GARDEN'

Devised by General Montgomery commanding the Allied 21st Army Group, Operation 'Market Garden' planned for airborne forces to seize bridges in advance of British 2nd Army as it struck north-east from the Meuse-Escaut Canal towards the Zuider Zee. Its aim was to cut off the German forces in Western Holland, out-flank the Siegfried Line defences to the north, cross the River Rhine and sweep down in the heart of Germany – to end the war in 1944. The essence of this daring and ambitious plan was *speed* to throw the enemy off-balance long enough to allow the Allies to complete their decisive stroke before being halted through lack of supplies. Two-pronged, the 'GARDEN' part of the operation was the ground advance of General Horrocks' British XXX Corps, with XII and XIII Corps advancing more slowly on the western and eastern flanks respectively. As the countryside was largely waterlogged, movement was virtually restricted to a single road, only wide enough for one tank.

The 'MARKET' part of the Operation required airborne troops to hold open the canal and river crossings on the Eindhoven-Arnhem road and defend them until relieved by ground forces, thus providing a 60 mile airborne carpet for the ground troops to advance over. This meant that five major bridges had to be taken and held, it was estimated that the northern-most bridge at Arnhem could not be reached until between 48 and 72 hours after the start of the operation. General Browning, the British Corps

Commander expressed a note of caution when he said 'We can hold the Arnhem bridge for four days, but I think we might be going a bridge too far'.

The operation was to be carried out by all available troops and aircraft of U.S. General Brereton's 1st Allied Airborne Army, under British General Browning's Corps command. There was the U.S.XVIII Airborne Corps (U.S.82nd and 101st Airborne Divisions); British 1st Airborne Corps (1st and 6th Airborne Divisions, 1st Special Air Service and 1st Polish Independent Parachute Brigades); the British 52nd Lowland Division – the Airborne Army's Air-Landing Formation; the Delivery Group, Engineers and an anti-aircraft unit – lifted by the IX United States Troop carrier Command and the R.A.F.'s 38th and 46th Groups. The entire airborne force for the 'Market' Operation consisted of more than 25,000 troops, 511 vehicles, 330 artillery pieces and 590 tons of equipment.

General Browning planned to bring about two-thirds of his Airborne Division in on Lift One on 17 September and to fly the balance of these divisions, with supplies, in on 18th and 19th; 52nd (Lowland) Division to standby for flying into an airfield north of Arnhem after it had been taken by ground forces. The Delivery Group, with engineers and anti-aircraft artillery, were to be held in readiness to glider-land in any of the divisional areas, to improvise and operate an airstrip, if necessary. From south to north, divisional tasks were:

1 *U.S. 101st Airborne Division* were to drop north of Eindhoven to seize and hold the bridges on XXX Corps line of advance – at Eindhoven, over the Wilhelmina Canal at Zon, the Zuit Willensvaart Canal at Veghel and the River Aa, plus two other smaller bridges.

2 *U.S. 82nd Division* were to drop and take the Groesbeek, the sole dominating land feature. They were also to take and hold bridges over the Maas at Grave, over the Maas-Waal Canal just west of Nijmegen and over the Waal on the northern outskirts of the town. This involved defending a perimeter much larger than could normally be held by a single division, subsequently Brown-

OPERATION "MARKET GARDEN"

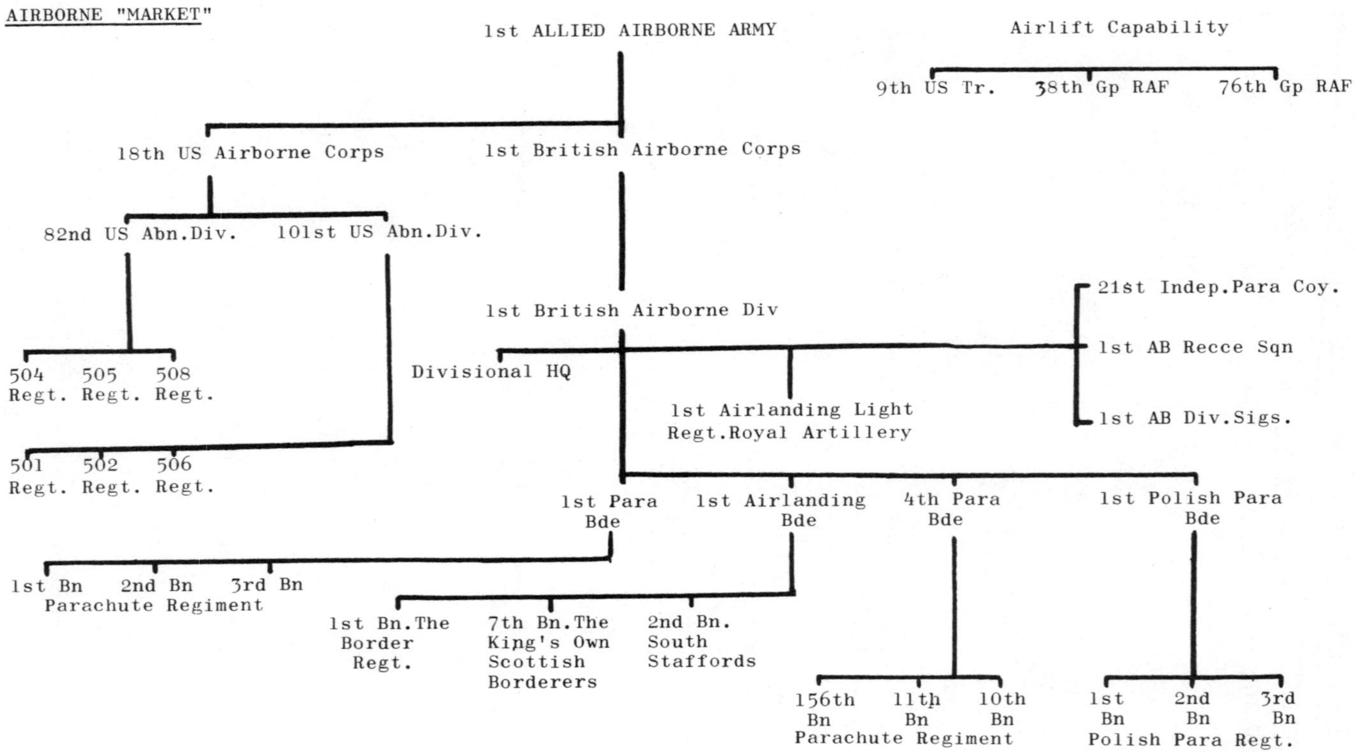

ing told General Gavin the Divisional Commander to first secure the high ground, leaving the Waal bridge until circumstances permitted dispersion of his forces. Lifts Two and Three were planned to arrive on the Groesbeek feature as reinforcements.

3 *1st British Airborne Division* Lathbury's 1st Parachute Brigade were to drop on D.Z. 'X' north of the river some six miles west of Arnhem bridge, which they were to seize, with the pontoon bridge 1200 yards away; hold north and south banks of the river east of Arnhem. Simultaneously, Hick's 1st Air-Landing Brigade in 345 Horsa and 13 Hamilcar gliders would land on L.Z. 'Z', just north of D.Z. 'X', secure the landing areas for the Lifts on the following day and then occupy an area on both banks of the river extending westwards from Arnhem about a mile beyond Oosterbeek. Gough's Reconnaissance Squadron (in armoured jeeps) were to make the initial assault on the Arnhem bridge. On the following day, Hackett's 4th Parachute Brigade were to drop from 127 C47s on D.Z. 'Y' 2,000 yards north-west of 'X' and 'Z'. L.Z. 'S' to the east of D.Z. 'Y' was to receive 286 Horsas and 15 Hamilcars carrying the remainder of 1st Air Landing Brigade and the Royal Artillery, who were to seize the high ground north of Arnhem. On the third day, Sosabowski's 1st Polish Independent Parachute Brigade Group were on-call to drop from 114 C47s on D.Z. 'K', one mile south of the Arnhem bridge, to assist in the capture of the bridge if necessary and pass through 1st Parachute Brigade to hold ground north-east of the town. More Polish troops and engineers were to come in gliders on L.Z. 'L', used on the previous day as a re-supply area.

The main tactical flaw in the British plan was the location of the dropping and landing zones for General Urquhart's 1st Airborne Division which ought to be dropped as close as possible to the Arnhem bridge, accepting some landing casualties to prevent the bridge being destroyed before capture. With 1st Parachute Bde. dropped close to the bridge and 4th Bde. on high ground north of Arnhem, supported and supplied from the air, the Division could have held out for a long period. But regarded as air operations, airborne activities were under R.A.F. control until the troops actually landed and concern over possible heavy casualties from

A.A. fire caused the R.A.F. to disagree to the proposed landing sites. Subsequently Generals Browning and Urquhart, led to believe by Intelligence Reports that their men could march to the objectives with little danger, agreed to dropping zones some five miles from their objectives in Arnhem.

The 82nd and 101st U.S. Divisions devoted more than three-quarters of their first-day lift to foot-soldiers, whereas the British decided to allot only half their first-day capacity to infantry, instead transporting all their vehicles and heavy gear on the first day rather than delivering all three parachute brigades supported by anti-tank weapons, or two parachute brigades and a glider brigade, keeping a parachute brigade in reserve. The 1st Air Landing Brigade had to remain out of the battle to guard the landing areas so that the 4th Parachute Brigade and the balance of the gliders could come in safely on the following day. Thus of four brigades, only one (the 1st) was to attempt to seize the Division's objectives during the first twenty-four hours.

In general, Browning's Corps plan possessed inflexible aspects that totally disregarded the enemy's reaction – a time-table laid down when and where each unit was to be delivered during the

The Arnhem Wargame. The Horsa, Hamilcar and Waco gliders of Hick's 1st Air Landing Brigade unload their occupants on Landing Zone "Z".

three days of the air lift, instead of the uncommitted troops being held in reserve to be employed as the situation demanded.

Optimistic appraisals of the operation arose from hopelessly incorrect Intelligence assessments of German strength in the area, and a disregard of Dutch Resistance reports of local concentrations of German armour (including '20 to 30 Tiger tanks'). In fact, General Bittrich's elite although battered veteran II SS Panzer Corps had reached rest areas in the vicinity of Arnhem a few days earlier. Extricated from battle and sent to the area for refitting and rehabilitation, Harzer's 9th 'Hohenstaufen' Division of 6,000 men, 20 Mk V tanks, some self-propelled guns, armoured cars and 40 armoured personnel carriers with heavy machine-guns plus artillery were hidden in the densely wooded National Park north and north-west of Arnhem. And Harmel's 10th 'Frundsberg' Division, desperately short of armoured vehi-

The Arnhem Wargame. Paratroopers of the 101 U.S. Parachute Division scattered over the area exactly as they landed.

cles but with formidable artillery, mortar and anti-aircraft units, was encamped in a semi-circle to the north-east, east and south-east of the Dutch town.

One of Germany's most competent, experienced senior officers, General-Feldmarschall Model had set up his headquarters in Oosterbeek, three miles from the broad expanse of heathland where the 1st British Airborne Division were scheduled to land on September 17th. To make room for Model's headquarters, Krafft's understrength SS Panzer Grenadier Training and Reserve Bn's three companies (another 1,000 SS recruits were due to arrive at any moment for training) bivouacked in the woods and farms north-west of Oosterbeek, not far from the village of Wolfheze – directly between the zones where the British 1st Airborne Division were to land, blocking the route into Arnhem.

Moreover, within the week preceding the attack, the experi-

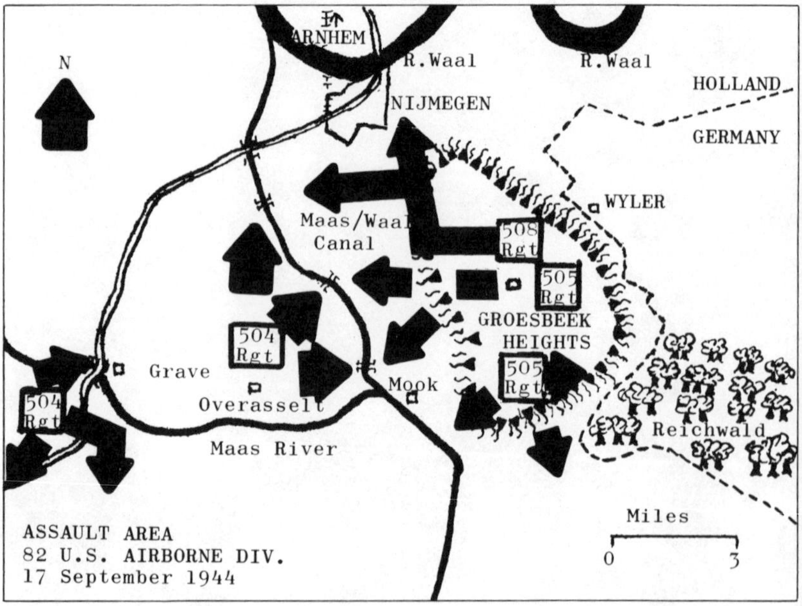

enced General Student had moved to an HQ at Vught, between Nijmegen and Eindhoven, to recreate his 1st Parachute Army and establish a defence-line in-depth behind the Albert Canal covering the front from Antwerp to Maastricht. One of Student's units, von der Heydte's 6th Parachute Regiment had provided stiff opposition to a Guard's Armoured Division attack on 10 September to secure a bridgehead across the Maas-Scheldt Canal twelve miles south of Eindhoven, as a preliminary to Operation 'Garden'.

General Model was an experienced soldier whose rapid reactions could well lead to improvised but flexible defences, and General Student, one of the most experienced airborne leaders in the world, could possibly out-guess the Allied commanders and frustrate their intentions.

During the night 16th/17th September, the R.A.F. attacked airfields in Holland; in the morning two waves of 139 Lancasters and 20 Mosquitos attacked German anti-aircraft positions, followed by 816 Flying Fortresses escorted by 161 Mustangs and 212 P47s, who bombed coastal anti-aircraft and inland batteries.

Sunday 17 September 1944 was bright and clear as Lift One took off from seven British and 17 United States airfields scattered from Dorset to Lincolnshire – an armada formed of two great streams of 2,023 carrier-planes, tugs and gliders. The smaller stream of 494 C47s and 70 Waco gliders carrying 101st U.S. Division flew a southerly course, crossing the English Channel at the North Foreland, proceeding almost due east to Gheel, wheeling left to dropping-and-landing-zones north of Eindhoven. A larger column, the U.S. 82nd Airborne and paratroop elements of the 1st British Airborne Division took off in 625 troop carrier planes and 50 gliders towed by C47s from airbases in the Grantham area. At five to 20 second intervals the IXth Troop Carrier Command's planes left the ground to rendezvous in wave after wave before setting out in three parallel streams to cross the coast near Alderburgh. From eight bases in Gloucestershire and Oxfordshire huge sky 'serials' raised gliders and tugs at an unparalleled launch rate of one combination per minute. Their very numbers made forming-up intricate and dangerous as they climbed slowly to altitude and headed west over the Bristol

Channel then, with their speeds synchronised, the tugs and gliders echeloned to the right in pairs, and made for the marshalling point above Hatfield,north of London. Innumerable Skytrains of British bombers – Halifax's, Stirlings and Albemarles, tugged equipment-and-troop-carrying gliders, bounding behind them at the end of 300 foot long ropes with massive Hamilcars ploughing along among the smaller Horsa and Waco gliders.

Landing furthest north at Arnhem, the British required artil-

lery and anti-tank guns in the first lift to capture and hold their objectives until land forces could link up. This meant that gliders carried the bulk of General Urquhart's Division. The 135 troop-carrying planes bearing Brigadier Lathbury's 1st Parachute Brigade did not take off until half an hour later because the paratroop-carrier planes were capable of 140 miles an hour against the 120 miles an hour of the unwieldly gliders and tugs.

Immaculate General 'Boy' Browning was in a glider piloted by

The Arnhem Wargame. Paratroopers of Gavin's 82nd U.S. Airborne Division, dropping right onto German anti-aircraft batteries, engaged the crews and put them out of action.

Brigadier Chatterton, Commanding Officer of the Glider Pilot Regiment, his Corps Headquarters bound for Nijmegen, travelled in a special series of 38 gliders, with the 82nd Airborne and 1st British Divisions along the northern track.

Air-Sea Rescue launches were positioned in a chain across the North Sea to pick up survivors on the few occasions when an aircraft put down with engine trouble, or a glider tow-rope snapped. Technical and other difficulties caused twelve gliders to ditch in the sea or force-land well short of their objective. Lift One was remarkably successful: enemy anti-aircraft fire destroyed 35 C47s and 16 Waco gliders carrying the American Divisions but not a single troop-carrier on the Arnhem lift was lost through enemy action. The extensive preliminary air operations were the reason for the slight losses in carrier aircraft during the outward journey, and the two carrier fleets were also protected by 371 Tempests, Spitfires and Mosquitos, 548 Thunderbolts, Mustangs and Lightnings.

Precisely at midday, in the face of light small-arms fire, Pathfinders of the 21st Independent Parachute Company dropped from twelve Stirlings to capture a German platoon position. On landing they set up their beacons to mark the L.Z.s and D.Z.s and waited for the deep throb of the glider-tugs and carrier aircraft.

The Arnhem Wargame. Men of the 506 U.S. Parachute Regiment attempt to assault over the blown bridge across the Wilhelmina Canal at Zon.

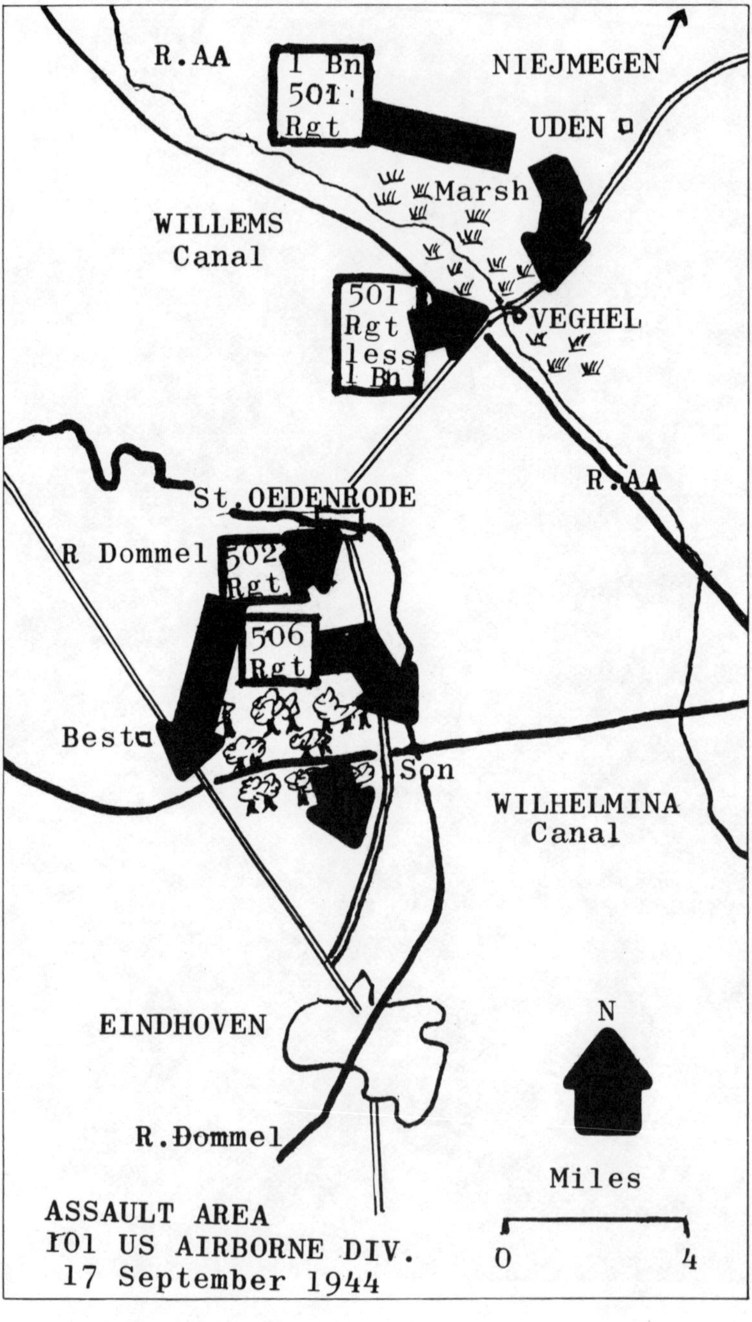

ASSAULT AREA
101 US AIRBORNE DIV.
17 September 1944

Occurring between 1315 and 1400 hours, the glider landings were the most successful of the war as the pilots skilfully brought in their flimsy craft on to the pathfinder's smoke signals and orange-and-crimson nylon markers. Two Hamilcars nosed into the soft soil and overturned, disabling two valuable heavy anti-tank guns, and 35 more gliders failed to make the landing zone, including those carrying the Reconnaissance Squadron's armoured jeeps. The glider-borne troops moved away as the C47s arrived and the blue sky blossomed with more than 2,000 varied coloured parachutes of the 1st Parachute Brigade floating gently down to earth, dropping accurately and safely onto their zone, with negligible landing casualties, and everyone in his right place at the right time.

Back in Belgium General Horrocks, standing on a slag heap by the Meuse-Escaut Canal, had his field glasses pointed northwards seeking the airborne armada. The Guards Armoured Division headed some 20,000 vehicles awaiting orders that would send them clattering down the road to Eindhoven. At 1330 hours, the tanks of the Irish Guards led off XXX Corps attack, behind a moving barrage fired by 350 guns; overhead 200 R.A.F. Typhoon fighter bombers screamed down to silence the German anti-tank batteries who, from well sited positions, were already inflicting losses on British armour. Immediately it became clear that the 'Garden' part of the operation was going to be a desperate and urgent business if the essential speed of advance was to be attained so that the Airborne divisions were not sacrificed.

Major Gough's Reconnaissance Squadron, without their jeeps, could make only a greatly reduced attempt through woods and villages to capture the Arnhem bridge by coup-de-main – none got through. Lieut. Colonel John Frost's 2nd Parachute Bn. marched in single lines on either side of the road that ran close to the north bank of the River Rhine, advanced into Heelsum and through the Doorwertsch wood. Fitch's 3rd Bn. north of them, set off on a more direct route along the main Utrecht-Arnhem road. Dobie's 1st Bn. moved off to approach Arnhem on the high ground via Wolfheze station on the Ede-Arnhem road. The peaceful atmosphere of the dropping zones prevailed for the first two miles of the advance, and well dressed Dutch civilians

poured out to greet their deliverers as they marched past red-tiled cottages and palatial villas.

The Air Landing Brigade took up defensive positions to protect the Landing Zone for the Second Lift's arrival on the following day. Forty officers and 700 other ranks of the 7th Bn. King's Own Scottish Borderers marched off to positions surrounding Dropping Zone 'Y' at Ginkel Heath, with pipers at their head playing 'Blue Bonnets'. It was to be the 7th KOSB's first and last action of World War II and at its end only four officers and 72 men remained. McCardie's 2nd South Staffords dug-in around the perimeter of LZ'S' near Reyerscamp, to the east of the Scots; Hadden's 1st Bn. Border Regiment occupied their allotted area south of the railway line at Renkum Heath.

The Germans reacted swiftly to the landings. Field-Marshal

The Arnhem Wargame. Paratroopers of 506 U.S. Parachute Regiment cross the engineers bridge over the Wilhelmina Canal at Zon on the first day of the battle.

The Arnhem Wargame. On the southern side of Nijmegen Bridge, it is debatable who is going to get the greatest surprise – the German tank crew or the careless loader of the American anti-tank rocket-firing bazooka who seems to be in a direct line with its recoil-blast!

Model abandoned his headquarters at Oosterbeek and drove to Arnhem where he gave orders to General Kussim, commander of the town garrison; then east to Doetinchem to Bittrich's 2nd Panzer Corps H.Q. who had already issued orders to his two SS Panzer Divisions.

General Horzer's 9th SS Panzer 'Hohenstaufen' Division were to rapidly occupy the Arnhem area and destroy the enemy forces who had landed to the west of the town, denying them the bridge at Arnhem. The 9th SS Panzer Division split into two groups on arriving in Arnhem on the evening of the 17th. Brinkmanns' group patrolled the town with armoured infantry to clear British parachutists from houses, while Spindler's Sperrgruppe (blocking-group) formed a barrier of armour and infantry on the outskirts of the town and patrolled forward along the three approach routes. The 10th SS Panzer 'Frunsberg' Division moved quickly to Nijmegen and occupied the main bridge area in strength.

A set of Allied operation orders found on the body of an

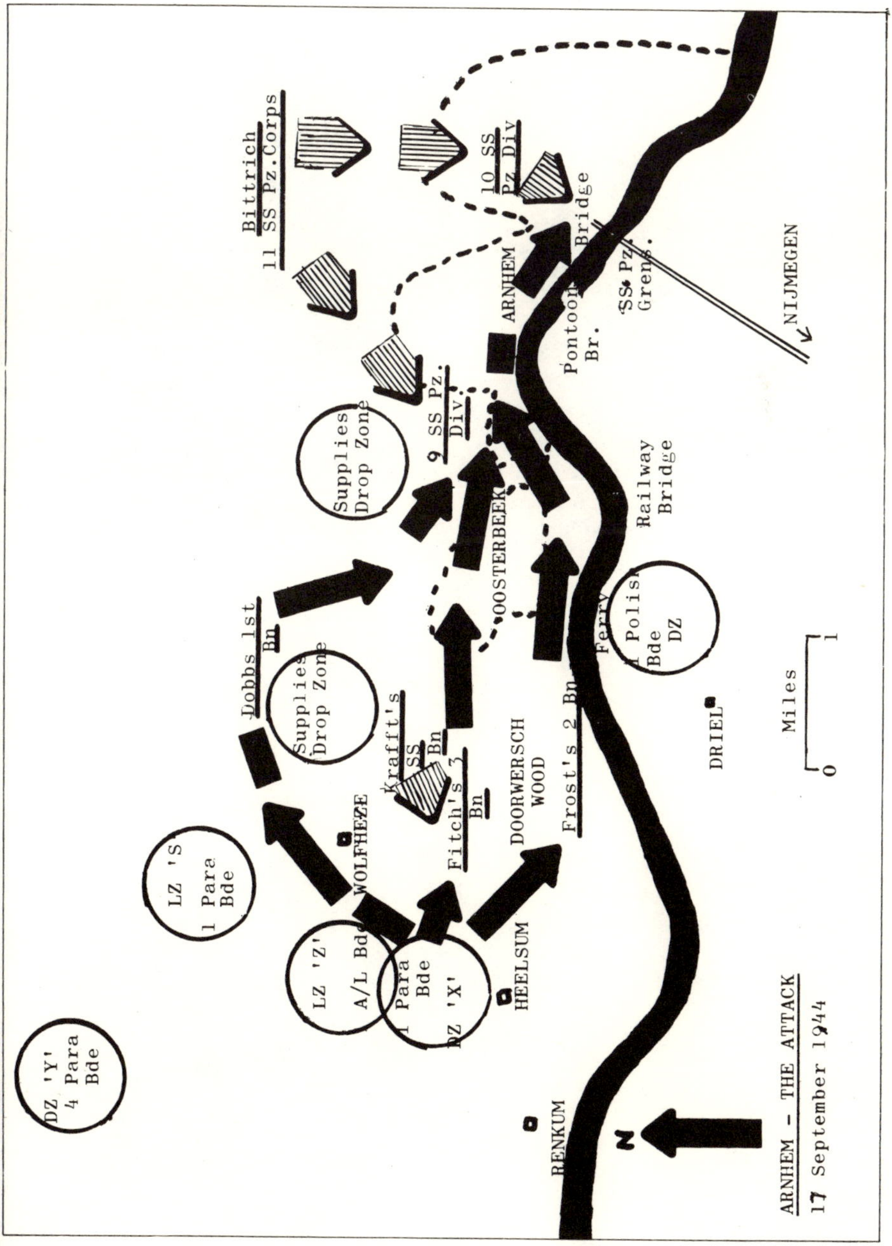
ARNHEM – THE ATTACK
17 September 1944
Bittrich
11 SS Pz.Corps
10 SS
Pz Div
Bridge
ARNHEM
SS Pz.
Grens.
NIJMEGEN
Pontoon
Br.
9 SS Pz.
Div.
Supplies
Drop Zone
Railway
Bridge
OOSTERBEEK
Ferry
1 Polish
Bde
DZ
Dobbs 1st
Bn
Supplies
Drop Zone
Krafft's
SS
Bn
Fitch's 3
Bn
DOORWERSCH
WOOD
Frost's 2 Bn
DRIEL
Miles
0
1
WOLFHEZE
LZ 'S'
1 Para
Bde
LZ 'Z'
A/L Bde
DZ 'X'
1 Para
Bde
HEELSUM
DZ 'Y'
4 Para
Bde
RENKUM
N

The Arnhem Wargame. 10th SS Panzer Division crossing the Rhine to the east of Arnhem, forced to do so because Frost's paratroopers were blocking the only bridge.

The Arnhem Wargame. German Panzer Grenadiers and light armoured forces attempt to rush Frost's defenders of the Arnhem Bridge.

American officer killed when one of the 82nd U.S. Division's Waco gliders was brought down by AA fire had come into General Student's hands, and after the war, the German Airborne Commander wrote: 'The importance of this capture was immense, for we learnt at once of the enemy's strength and intentions and the speed and comparative success of our counter-action was to no small extent due to early knowledge of this hostile move'.

At the time of the drops, the only Germans between the landing areas and Arnhem were the SS Training and Depot Bn. under Krafft who witnessed the landings and positioned his force in a screen running north to south, about halfway between Oosterbeek and the British assembly areas. Acting as a delaying group between the D.Z.s and the bridge, Krafft's force won time for the 9th SS Panzer Division to get into position, and played an important role in the battle that had just begun. As the day drew on, in the centre Lippert and Krafft's battalions were in position with

The Arnhem Wargame. Frost's defending force behind their barricade.

The Arnhem Wargame. View from behind the barricade on Arnhem Bridge as Frost's anti-tank gunners and paratroopers fight off the German Light Armour and Panzer Grenadiers.

The Arnhem Wargame. The German attacker's view on the Arnhem Bridge.

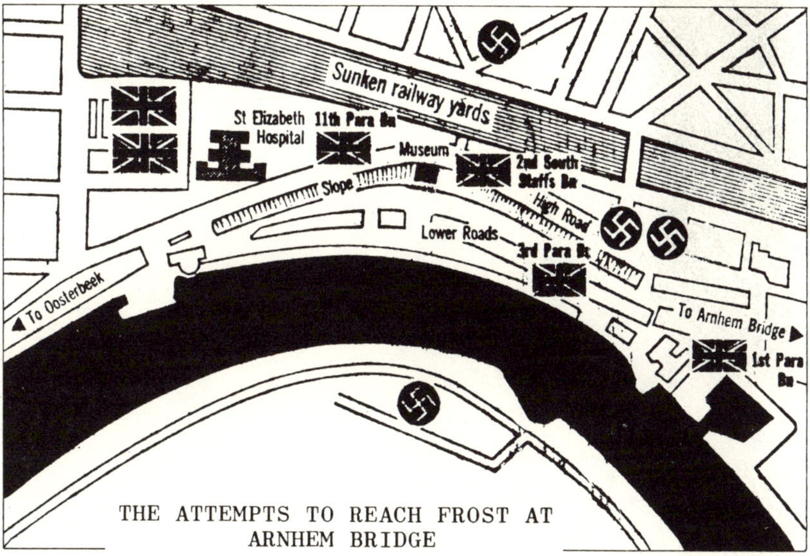

THE ATTEMPTS TO REACH FROST AT ARNHEM BRIDGE

machine-guns, mortars and riflemen; von Tettau's units were moving in from the north and west; and Brinkmann's men were edging along the north bank of the river.

Through the mist of the Autumn evening, the British 1st Airborne Division and the German 9th SS Panzer Division were both moving towards the Arnhem bridge, with the Germans possessing the great advantage of being aware of their opponents general position and intention, whereas the Airborne force were not even aware of the existence of the German armoured formation!

Meanwhile, the 1st Bn. on the river road had advanced through Heelsum, ambushing German vehicles and taking prisoners despite opposition in the Doorwertch wood. B Company attacked enemy machine-gun positions on the Den Brink high ground and then moved to capture the Brigade's second bridge target, the pontoon bridge that was still undamaged although the Germans had removed some of the barges and towed them to a nearby dock. Throughout, the 1st could raise nobody on their radios and inter-unit communication was non-existent.

By dusk, A Company, 2nd Bn. were near the road bridge in Arnhem and Lieutenant Grayburn's platoon attempted to rush

The Arnhem Wargame. View of Frost's force defending the houses around the bridge approach (at bottom right-hand corner).

The Arnhem Wargame. German armour and infantry close in on Frost's force defending the houses around the northern end of Arnhem Bridge.

the bridge from the north but were turned back by two quick-firing 20 mm flak-guns and the machine-guns of an armoured car. Although wounded, Grayburn organised his force in a house near the approaches to the bridge.

Frost led his determined party towards their objective; taking advantage of the darkness which hampered German fire and hindered their armour, the paratroopers brushed aside minor opposition at the northern end of the road bridge which they reached at 2000 hours. Frost established his headquarters in a house in Ne Kraan street north-west of the bridge, which was intact but under mortar fire, and sent a runner to B Company ordering them to cross the river in boats to secure the southern end of the bridge. But this group was still engaged in a sharp fight at Den Brink and there was no sign of C Company, who were fighting hard at the railway bridge near Den Brink. Fortifying the houses controlling the northern approaches to the bridge, Frost's party launched several attacks against the Germans holding the southern end of the objective, but all were beaten back.

During the night elements of Fitch's 3rd Bn., 1st Brigade Headquarters, some Engineers and the headquarters of the Reconnaissance Squadron, with a platoon of service troops slipped through on the same southern river road to reinforce Frost's party, so that by daylight he commanded some 600-700 men. This small party represented the net strength on the main objective from a tactical plan based on more than 10,000 men, 92 guns, 500 jeeps, 400 trailers and 300 motor-cycles.

Their approach route remained open for so long that better reconnaissance could have re-directed the whole of the 1st Parachute Brigade to the bridge by that route. Had not the 1st Air Landing Brigade been ordered to protect D.Z.s and L.Z.s for Lift Two's arrival on the following day, they could have been pushed through hard on the heels of Frost's party. Nevertheless this small force on the bridge had completely disrupted German defensive plans, as only one company of Panzer Grenadier Infantry had crossed the bridge to reinforce the Nijmegen garrison before Frost's party had isolated the main body of 10th Panzer Division on the wrong side of the Rhine. Model must have been well aware that if the airborne attack at Nijmegen were to be held

The Arnhem Wargame. British paratroopers man the houses and barricaded road at the northern end of Arnhem Bridge.

The Arnhem Wargame. The 17pdr anti-tank gun and paratroop machine-gun team man the barricade on the road leading to the Arnhem Bridge approach.

The Arnhem Wargame. Paratroopers of Frost's 2nd Parachute Battalion, including mortar and machine-gun teams, defend the houses at the northern end of Arnhem Bridge.

and the XXX Corps relief force halted, then the small party of British paratroopers on the northern edge of the Arnhem bridge had to be rapidly removed.

On the central approach route, the 3rd Parachute Bn. had been slowed by mortar fire from Krafft's defensive screen and then halted completely by armoured vehicles of 9th SS Division. In an early encounter General Kussin, Commander of the Arnhem garrison, drove headlong into the battalion and was immediately killed.

The 1st Bn. on the northern approach route were warned by Major Gough that tanks were blocking the road and Panzer Grenadiers in position along the railway line from Wolfheze to Arnhem; subsequently the battalion entrenched themselves on the edge of the line of woods east of Wolfheze station.

General Urquhart, badly handicapped by lack of communica-

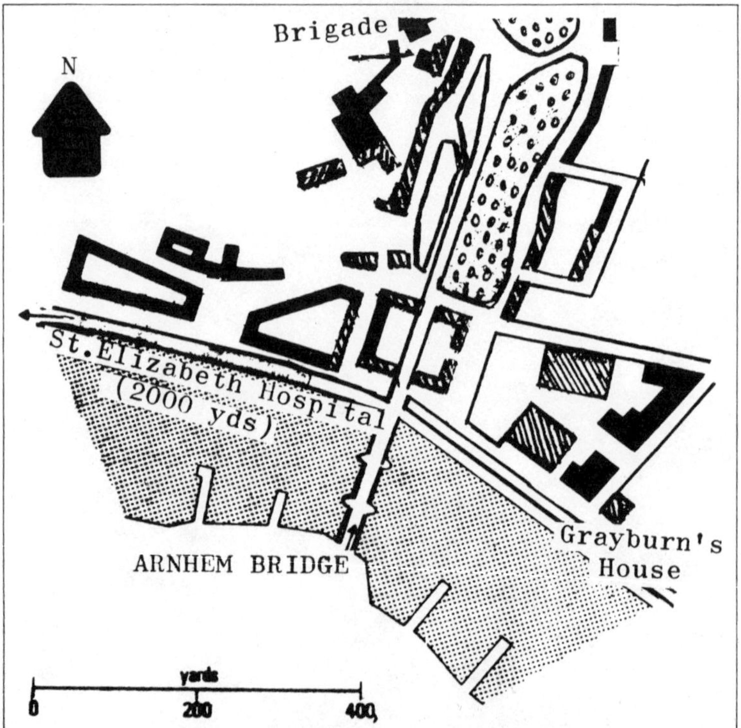

FROST'S DEFENSIVE POSITIONS AT THE BRIDGE

tion through inefficient radios whose effectiveness was limited by the heavily wooded terrain, had to personally acquaint himself with the situation so he and Brigadier Lathbury set off in a jepp and eventually spent the night of the 17th with the 3rd Bn. halted on the Utrecht- Arnhem road.

Throughout the night the struggle for Arnhem bridge swung in the balance and substantial reinforcements might have enabled Frost's party to decisively affect the battle. By daylight they were securely lodged at the northern end of the bridge in groups in houses and warehouses, repulsing with grenades, anti-tank guns and Piats, attacks by armoured cars and Panzer Grenadiers in half-tracks. German counter-attacks were prevented from forming-up on the southern approaches by continuous machine-gun fire directed across the bridge.Soon the northern end, under fire by guns and mortars, was a blazing inferno strewn with the wreckage of vehicles and the debris of war.

Gradually, British 1st Airborne Division were relinquishing the initiative, being unable to hold off assault by mobile armoured forces because of lack of offensive air-support fire-power denied them through confusion in responsibility.

In a house outside the main defensive perimeter, exposed and difficult to defend, Lieutenant Grayburn and the survivors of his platoon held out against ceaseless enemy attacks for two days, until German tanks and self-propelled guns firing at under 100 yards range forced them from their positions. Grayburn led repeated fighting patrols to prevent the enemy laying demolition charges, consistently exposing himself to point-blank enemy fire until he was killed by a flame-thrower. Lieutenant Grayburn, 2nd Parachute Bn., was posthumously awarded the Victoria Cross.

Lack of communications now played a vital part in the battle. On the morning of the 18th, 1st Parachute Brigade plus divisional troops were to be dropped on the western D.Z.s, held by the Air Landing Brigade. But the Germans had successfully halted the advance from these D.Z.s, and to use them again was to reinforce failure. Had 4th Brigade's D.Z. been switched to the subsequently unused D.Z. 'K' *south* of the approaches to Arnhem bridge, then they could have linked-up with Frost's force AND

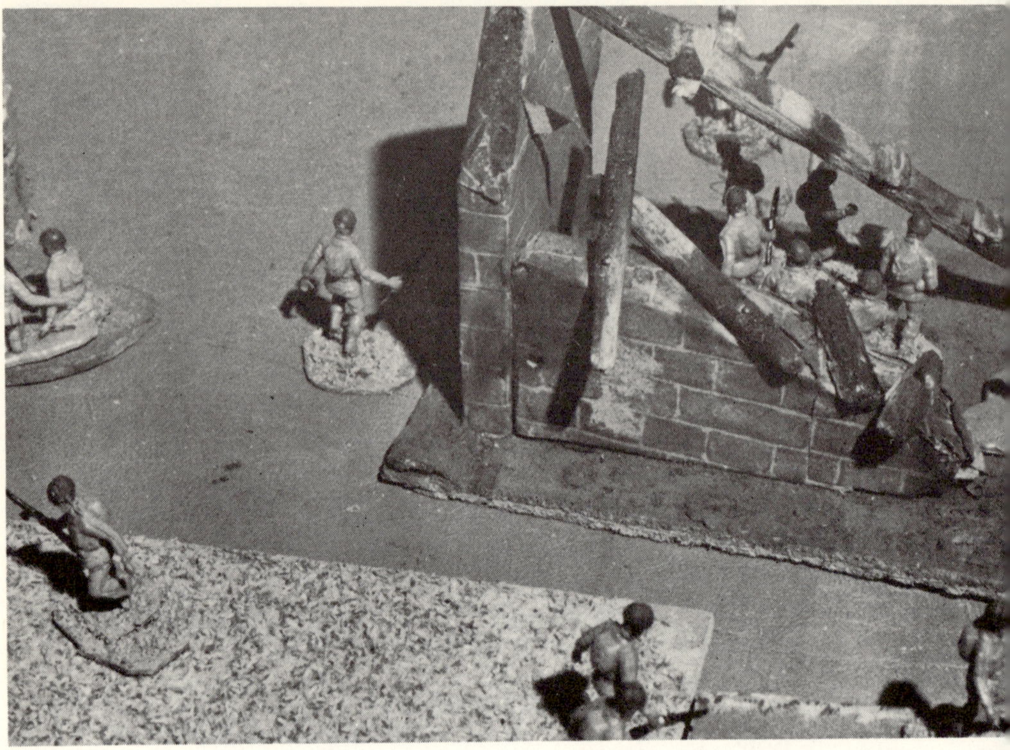

The Arnhem Wargame. Mortar teams and parachutists of Frost's 2nd Parachute Battalion defend wrecked houses at the northern end of Arnhem Bridge.

The Arnhem Wargame. A German Mark IV tank with a Panther in the background cautiously noses its way between the ruined houses defended by British paratroopers at the Arnhem Bridge.

The Arnhem Wargame. Amid the ruined houses around the approach to the Arnhem Bridge, a German Mark IV tank noses its way warily forward.

The Arnhem Wargame. Heavy German armour and supporting Panzer Grenadiers in the attack on Frost's 2nd Battalion defending the Arnhem Bridge approach. The PIAT man hidden in the ruined house might well make his mark on the seemingly unsuspecting closed-down Panther!

the three battalions of the Air Landing Brigade would have been freed from their unproductive task to take a more constructive part in the battle. With General Urquhart missing, no one on-the-spot could alter arrangements and General Browning, on the Groesbeek, could have no idea of the situation.

Fog delayed landings until 1500 hrs, when Hackett's 4th Parachute Brigade were dropped and decimated four miles west of their objective, being flown in at 500 feet through A.A.shell-bursts and the smoke from blazing woodland around the zones, dropping into heavy small-arms fire. The Air Landing Brigade were fighting hard to keep the enemy off the D.Z. – Payton-Reid's 7th KSOB fixed bayonets and desperately charged to clear the woods, fields and ditches of Germans firing at the descending parachutists. On landing, Brigadier Hackett learned that General Urquhart was missing and that Brigadier Hicks of the Air Landing Brigade was acting in his place (as the General had earlier specified). Being senior to Hicks, Hackett drove to Divisional Headquarters where the two Brigadiers had an argument as to

The Arnhem Wargame. A German Mark III supported by Panzer Grenadiers moves forward against the British paratroopers defending the houses around Arnhem Bridge.

The Arnhem Wargame. A powerful Jagdpanther, supported by Panzer Grenadiers, grinds slowly forward down the road leading to the parachutist-defended northern end of Arnhem Bridge.

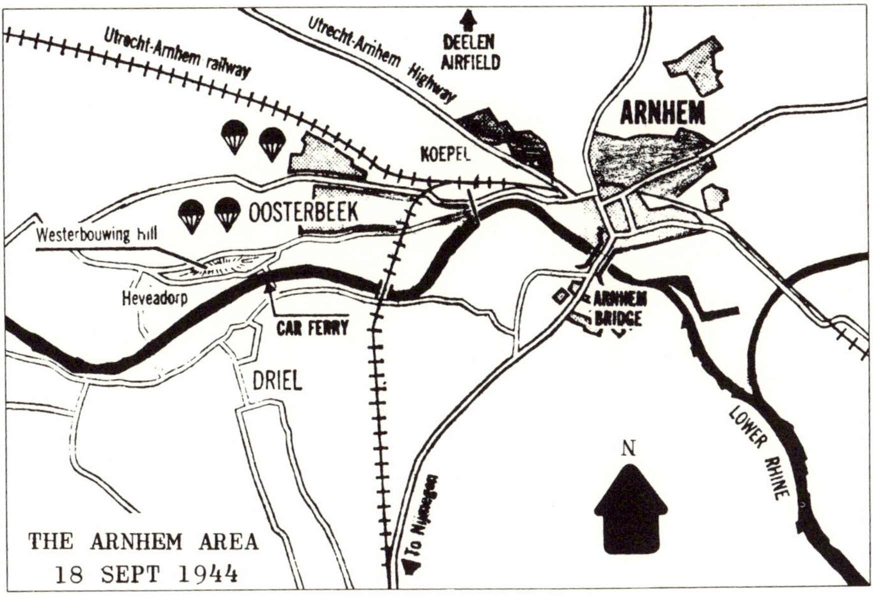

THE ARNHEM AREA
18 SEPT 1944

The Arnhem Wargame. German Panzer Grenadiers supported by Panther and Mark III tanks move forward against the 2nd Parachute Battalion, defenders of the approaches to Arnhem Bridge.

who was in command!

The strongest efforts were being made to reach Frost. Before nightfall on the 17th, Fitch's 3rd Bn., enfiladed from high ground on both sides by machine-guns and mortars, and by artillery firing from the south of the river, desperately tried to press forward into the town. Eventually the regiment were split in half and fell back under cover of darkness to the Rhine Pavilion on the embankment. At daylight on the 18th, the 3rd Bn., with the 1st and 11th Bns. and the South Staffs, doggedly fought their way yard-by-yard into the town, trying to force a passage through a network of streets lying between the railway line and the river. The 1st Bn. moved along the lower road on the embankment, swinging left for the bridge, as the other two battalions advanced along the main road from the St. Elizabeth hospital. The South Staffs battalion front was only the width of the street, and from the upstairs rooms of the houses Germans fired machine-guns and dropped grenades onto their heads, causing them to take shelter in a museum until forced out by mortars, 20 mm guns and

The withdrawal to the Oosterbeek perimeter on 20th September.

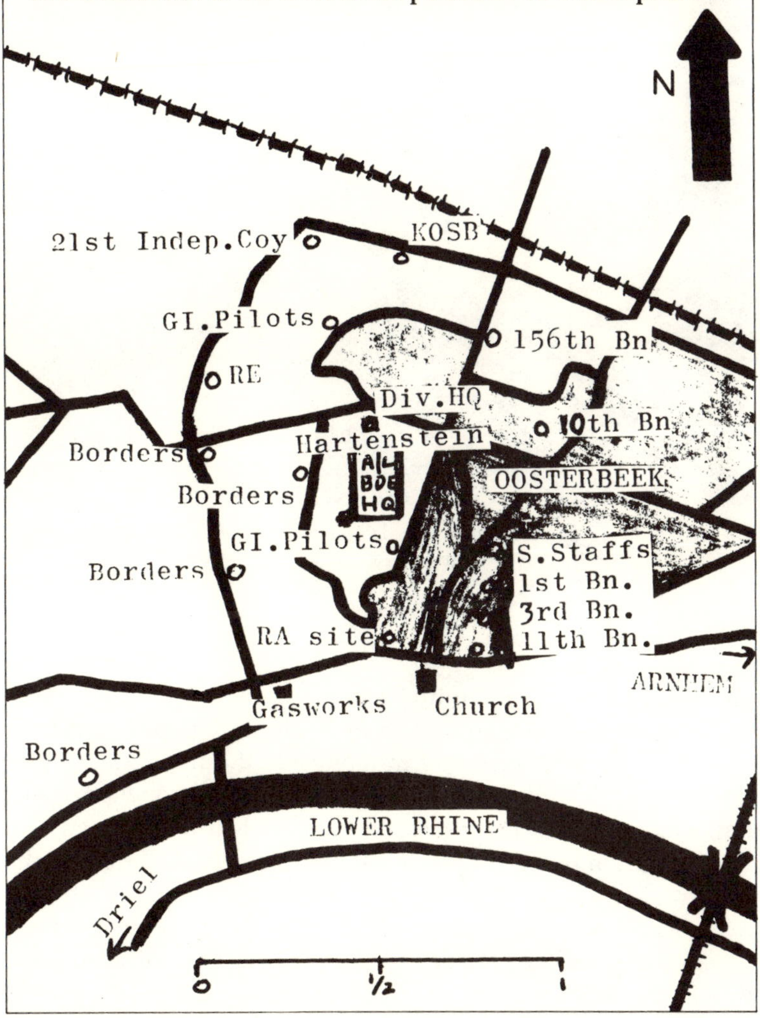

self-propelled artillery. Then tanks swept down the road to the hospital, wiping out the South Staffs and the 11th Bn. until only a small group of survivors escaped to the outskirts of the town. By 0700 hours, Dobie's 1st Bn. with only 49 men left standing, had virtually ceased to exist. Dobie was wounded and taken prisoner, Fitch was killed by a mortar bomb.

Hackett's 4th Brigade, led by the 156th Para.Bn., moved

The Arnhem Wargame. A Sherman and Churchill tank of the Guards Armoured Division nose cautiously through the yard of a Dutch farm.

along the line of the railway past the Reyerscamp L.Z., where the gliders had brought in the 4th Brigade's transport and equipment towards the high ground north of Arnhem, but were repulsed with heavy losses, and dug in alongside the main road under devastating fire from anti-aircraft guns, self-propelled guns, tanks and mortars.Their task on Ginkel Heath concluded, 7th KSOB were moving on the left flank to secure L.Z. 'L' for the Polish gliders landing on the following day, while the 1st Borders, who had taken heavy casualties at Renkum Heath, were advancing south of the railway line towards their allotted position in the Arnhem perimeter. These battalions were supported by Glider Pilots and other divisional units acting as infantry, and three 75 mm batteries of Johnson's 1st Light Regiment R.A. But by evening, 4th Parachute Brigade was beginning to share the fate of the 1st, as the new arrivals vainly but bravely attempted to pierce the German armoured screen that stood between them and 2nd Parachute Bn., still grimly holding on to their end of the bridge, hoping that reinforcements would soon arrive.

Attempting to return over the Arnhem bridge early in the morning of the 18th, the SS Reconnaissance Unit were engaged by every weapon, including anti-tank guns, Piats and grenades, until they turned back leaving eleven blazing armoured cars and half-tracks on the embankment.

Frost's original force (A and C Coys. of the 2nd Bn., C Coy. of the 3rd Bn.;1st Parachute Brigade H.Q.; elements of the Recce squadron led by Major Gough and a few anti-tank gunners, engineers and RASC (personnel) were now reduced too heavily to make any further attacks on the southern end of the bridge, and were garrisoning some 40 houses and a school. The area was blazing fiercely, and Tiger tanks rumbled through the streets blasting buildings at point-blank range and forcing the defenders to evacuate them one by one. Frost was wounded and Major Gough assumed command of the rapidly diminishing force who continued to offer resistance, securing minor successes against tanks with their anti-tank guns and Piats.

Both British and Germans were discovering that street fighting absorbs soldiers in the same way as blotting paper absorbs ink, and a battalion spread over a map-square (a kilometre by a kilometre) are hard put to man a cohesive defensive line whereas the same force in open country could occupy and hold the area. It is even more marked in the attack because the objectives have to be cleared at every level which multiplies as much as tenfold in the manpower requirements. Street fighting also absorbs time as each action has to be repeated at several levels.

On the following day the 19th, the Polish Brigade were due to be flown in on Lift Three landing south of Arnhem bridge. Adverse weather conditions forced their flight to be postponed for twenty-four hours; their zone was changed and their gliders were scheduled to come in on an L.Z. north of Oosterbeek. But this area was not securely held by British troops, nor was the Supply-Drop Point 'V' north of Warnsborn, where 38th and 46th Groups R.A.F.were scheduled to make a supply drop on the afternoon of the 19th. A signal sent requesting a change of dropping zones was not received, causing both drops to be tragically wasteful.

The 163 supply aircraft came in at 1500 feet into murderous anti-aircraft fire that flayed them for eight minutes as they slowly crossed and re-crossed the dropping zone while despatchers

The Arnhem Wargame. Sherman tanks of the Guards Armoured Division cross the engineer-erected bridge over the Wilhelmina Canal at Best some miles west of Zon, in their attempt to reach the airborne forces fighting ahead of them.

desperately pushed out 190 tons of food and ammunition – mostly to the waiting Germans. Those British paratroopers who could see the drop were wild with anger and frustration, leaping from slit-trenches as they tried desperately but unavailingly to attract the pilots attention. Fourteen aircraft were shot down and 97 damaged in this courageous but futile action. The tug-planes and gliders carrying the Poles located the landing-zone, and the 31 gliders remaining out of the original 46 met an inferno of fire as a squadron of Messerschmitt ME109s decended on the helpless gliders, riddling their thin canvas and plywood. Some broke up in the air or caught fire as the petrol in the punctured tanks of the jeeps they carried set them alight. Then the aircraft were gone and torrents of AA fire hit them. Gliders, some on fire and others badly damaged, crash-landed, ploughing into fields and trees, landing directly in the middle of a battle. In the confusion the Poles took fire from both friend and foe – and returned it. Then, under heavy fire, they blew off the glider tails with explosive charges and unloaded their equipment. Trailers and 6 pdr anti-tank guns were hitched up to jeeps and, running a gauntlet of fire three of the eight guns reached British lines. Many of the Poles, bewildered and shocked, were taken prisoner.

On the 19th General Urquhart managed to re-joun his force at Oosterbeek, having been trapped with Brigadier Lathbury in a house in Arnhem surrounded by Germans. Realising that the link-up with Frost's bridge defenders was out of the question, he now had to plan to survive while maintaining a bridgehead-reception zone for the 2nd Army when they arrive on the north bank of the river. He proposed to defend a four mile square perimeter based on Oosterbeek with its southern edge resting along the bank of the Rhine for about a mile; his Divisional headquarters were near its centre at the Hartenstein Hotel. At the exteme south-western corner of the perimeter, weakly defended and vulnerable to German attack, was the Heveadorp ferry which, firmly held on the northern bank, could provide Horrocks with a reasonable alternative to the intact Arnhem road bridge. This might also have occurred to General Browning who switched the long delayed drop of the Polish Parachute Brigade

The Arnhem Wargame. The 504 U.S. Parachute Regiment crossing the Waal in assault boats.

to a D.Z. at Driel south of the ferry.

Resembling the thumb of the right hand jutting northwards from the river bank, the perimeter was garrisoned by the 21st Independent Parachute Coy. in the north-west corner; on the left were about 250 men of the 7th KOSB defending the White House Hotel, reduced to rubble and the scene of fierce hand-to-hand fighting; behind were Mackenzie's Airborne Light Artillery Regiment of 75 mm Howitzers; the north-eastern corner of the perimeter was defended by the 156th and 10th Bns. of 4th Parachute Brigade; the western half manned by three skeleton companies of the 1st Borders, the Polish Glider troops and some Engineers; in the south-east near Oosterbeek Church were a mixed force of about 400 survivors of the 1st, 3rd and 11th Bns. of the Parachute Regiment and the South Staffords under Major Lonsdale; the Glider Pilot Regiment held two positions, one at the artillery site and one in a wood further north.

Organised resistance at the Arnhem bridge ended on the morning of Thursday 21 September, after Frost's small force had

The Arnhem Wargame. Polish glider-born troops, landing in the middle of the battle on their LZ north of Oosterbeek, attempt to hitch-up anti-tank guns brought in by glider.

achieved the whole Division's objective by holding the bridge against overwhelming strength for three and a half days. Conditions had steadily deteriorated until the only means of moving under cover was by blowing holes in the dividing walls of houses. Such vital defence buildings as the school were in ruins, walls so perforated as to be no longer bullet-proof and rubble piled high on all sides. The defenders, mostly wounded, huddled in twos and threes manning positions that really required twice their numbers but, believing themselves superior to the enemy of whom they had killed four times their own number, their morale was still high. At dusk on the 20th and throughout the night the last remaining strongholds were being assaulted by tanks, self-propelled guns and flamethrowers. Many attacks were repulsed but it had become hopeless and by morning when half Frost's force were casualties, 200 wounded including Frost himself surrendered to the Germans; most of the rest of his force were captured trying to make a fighting withdrawal.

Whilst all this had been occurring in the British area of operations, the two U.S. Airborne Divisions had been resolutely fighting and achieving their objectives to the south of the Rhine. On the 17th, 101 U.S.Division's pathfinders had quickly marked the D.Z.s and L.Z. near St. Oedenrode and Veghel, so that the division made a well patterned drop according to plan, supported by fighters and dive-bombers attacking anti-aircraft emplacements and engaging eight German tanks near the main dropping zone, destroying two and driving off the remainder. Otherwise no significant opposition was encountered on the ground and the units were able to assemble quickly and move out to secure their objectives. Reaching the bridge on the southern edge of Zon, 506th Parachute Regiment found it had been blown. Engineers rapidly set up a replacement and by midnight the regiment was south of the Wilhelmina Canal. On the following morning they encountered fierce resistance on the northern outskirts of Eindhoven, but by midday had outflanked the German positions, entered the town and made contact with the leading Reconnaissance elements of the Guards Armoured Division.

Further north, the main body of 502 Parachute Regiment secured the divisional bridgehead while part of the formation

The Arnhem Wargame. A general view of the eastern side of the British perimeter at Oosterbeek, the wall of the church bordering the road on the left.

captured St Oedenrode and its bridge intact before dark. Another detachment, after first being repulsed, seized the bridge over the Wilhelmina Canal at Best, some miles west of Zon. By dusk on the following day, the Guards tanks had crossed this canal on an engineer-erected bridge and were through St Oedenrode and on their way to Veghel, which had been the objective of 501 Parachute Regiment, dropped on both sides of water obstacles to directly attack and seize two roads and two rail bridges. By 1500 hours on the first day the Regiment had taken its objectives and were well dug in and ready to resist counter-attacks but encountered little opposition, and were at full strength and fresh when British armour passed through Veghel at 0645 hours on the 19 September. The 101 Airborne Division had successfully accomplished their part in the 'Market' part of the operation.

Gavin's 82nd U.S. Airborne Division, dropping only ten minutes after their Pathfinders, encountered considerable anti-aircraft fire around their dropping-zones and lost a number of C47s. However, in some areas paratroopers dropped right onto anti-aircraft batteries and immediately put them out of action.

Tucker's 504th Parachute Regiment set off in four directions to

The Arnhem Wargame. The northern aspect of the defensive perimeter at Oosterbeek.

secure bridges over the Maas-Waal Canal and the vital Maas bridge north of Grave, where a company had dropped in a direct attack. One platoon fell close to the bridge and moved towards the canal under enemy fire from buildings and a flak-tower on the southern approach to the bridge. The American paratroopers knocked it out with bazooka fire and turned its 40 mm gun against the enemy. The platoon was joined at the bridge by the 2nd Bn. and by dusk 504 regiment were holding the Grave Bridge and the southernmost of the four bridges over the canal. General Gavin himself described how it was captured:

'By keeping the bridge under fire and slowly walking in, the leading unit finally managed to drive out the German defenders into a house, on a small island on the locks of the canal. Then by keeping them pinned down by firing into the house, the troopers managed to get onto the bridge and cut the wires and remove the charges'.

Two central bridges over the canal were blown before paratroopers could reach them and the northern one, at Honinghutie, was the objective of another regiment.

Ekman's 505 Parachute Regiment captured Groesbeek and organised it for defence and, although enemy probing attacks

The Arnhem Wargame. The western defensive positions on the perimeter at Oosterbeek.

were seen building up, they had not been attacked by nightfall.

The most easterly D.Z. just south of Wyler was to be used as L.Z. for Second Lift gliders on the following day, so it had to be defended by Lindquist's 508th Parachute Regiment, who had jumped onto it. Also, they had to defend the northern approaches to the Groesbeek Heights, assist in capturing Honinghutie road bridge on the main XXX Corps axis. Finally, they were to capture the Nijmegen bridge, but only if it did not seriously weaken the division's hold on the heights. In the hope of securing the bridge intact, General Gavin ordered Lindquist to send a battalion under cover of darkness to achieve the objective. The chosen battalion, Warren's 1st, penetrated about halfway through the town of Nijmegen before being forced to take cover in houses – in the post office they destroyed the electrical controls for blowing the bridge.

The first night of the battle found the 82nd U.S. Airborne Division more or less intact and in full control of the dominating ground; they had captured the Grave Bridge, a bridge over the canal (although not the one on the main axis) and had prevented the Germans destroying Nijmegen Bridge.

The 18 September was a day of hard fighting for Gavin's 82nd

THE SHRUNKEN PERIMETER AT ARNHEM
IMMEDIATELY PRIOR TO EVACUATION

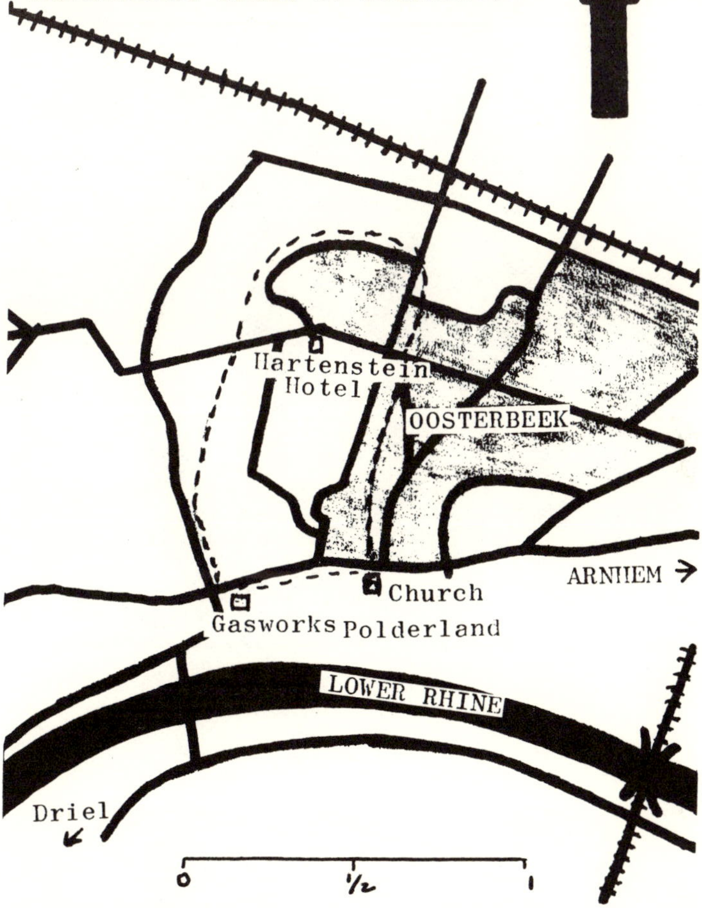

U.S. Airborne Division. Battling desperately for the Honinghutie Bridge, the 508 Regiment asked the 504th Regiment to try and attack the German defenders in the rear. With the Germans fully occupied fighting to the east, a 504th patrol crept across the bridge and fired into their rear, causing their resistance to collapse and although damaged, the bridge was secured.

The 1st Bn. 508th Regiment had to disengage from Nijmegen to join the main body of their regiment in a fierce battle to eject Germans who had overrun the landing-zones on which the 82nd's gliders were due to arrive. Fortuitously, they had been

The Arnhem Wargame. British paratroopers and 6pdr anti-tank gun in hastily dug defenses on the perimeter at Oosterbeek.

The Arnhem Wargame. The 6pdr and 17pdr anti-tank guns dug-in on the Oosterbeek perimeter defence position.

delayed two hours, even so the gliders landed under fire, but the delay destroyed any chance of capturing Nijmegen Bridge on that day. This lift consisted of 450 Wacos carrying three light field artillery bns., the balance of the anti-tank bn. and some additional divisional troops – all were in action by mid-afternoon.

At 0830 hours on 19 September, reconnaissance elements of XXX Corps contacted the 82nd U.S. Division's road block south of Grave and the main body of the Guards Armoured Division started coming on the scene at noon. General Gavin committed Vandavort's 2nd Bn. 505th Regiment, assisted by the Grenadier Guards Tank Bn., to capture the Nijmegen Bridge and by nightfall their forces were within sight of the southern end of the bridge. Then the attack ground to a halt short of the objective, largely because of lack of sufficient infantry to press it home – the same weather conditions that had held up the Polish brigade had also grounded Gavin's Glider Regiment, nor could he send reinforcements because everyone was involved in repulsing German attacks against the Groesbeek.

On this day the Germans, unable to reinforce their defensive positions holding the Waal rail line while Frost still blocked the Arnhem Bridge, crossed the Rhine well to the east. The 10th SS Panzer Division ferried tanks and vehicles across the river in what was a slow business, although apparently undetected by the 2,000 Allied fighters and bombers supporting the Market' operation.

Meanwhile, on the evening of 20 September, Generals Gavin and Adair, commander Guards Armoured Division, made a plan for an attack on the following morning, when an American parachute regiment were to cross the Waal by boat, followed by a simultaneous assault against both ends of the bridge. Assault boats were to be brought up on XXX Corps transport, but the attack although supported by fire from tanks and artillery and with smoke concealing the crossing, would still be exceedingly hazardous in the face of a determined German defence.

First, Nijmegen had to be cleared of German forces to allow the assault troops to gain access to the south bank of the Waal, a mopping-up process that took the whole of Wednesday morning. It was not until 1500 hours that the 504th was in position to

The Arnhem Wargame. View of part of the Oosterbeek defensive perimeter from the tower of the Church.

launch their assault boat into the 400 yard wide fast-flowing river. The smoke cover was not very effective and only half the first wave of boats reached the north bank, many being destroyed by enemy fire and swept away by the strong current. However, some 200 determined men scrambled ashore and established a shallow bridgehead, being gradually reinforced by further waves of boats crossing the bullet-swept swirling water in broad daylight. By 1830 hours the Americans had routed enemy opposition to the bridgehead and were moving towards the road bridge; it is said that they signalled success by raising the Stars and Stripes on the northern end of the railway bridge.

On the southern side of the bridge, the Guards Armoured Division had been fighting their way through, supported by Vandervoort's 2nd Bn. 505th Regiment, whose mortars and artillery pounded the German defences while men advanced from house to house. The Guard's tanks, including Sherman Fireflys with 17 pdrs forced their way to the southern approaches as the enemy defence ring of anti-tank guns were knocked out one by one. Finally there were only four self-propelled guns dug into the centre of a traffic circle, and at 1600 hours they were overrun by

American paratroopers with bayonets and grenades. Then, British tanks, lined up four abreast, charged through the little park that led to the open approaches to the apparently intact great Waal Bridge. It was later revealed that the Germans attempted to blow the bridge while attacking tanks were crossing but the explosives failed to detonate. A troop of four Guards tanks in line ahead charged across the bridge, coming under fire from an 88 mm a/tank gun sited on the other side of the river – 100 yards from the north end of the bridge in a sandbag emplacement by the side of the road. One of the four tanks and the 88 exchanged four rounds apiece, as the tanks machine-guns chattering away; suddenly the big German gun blew up. Clinging to the girders of the bridge, Germans with grenades, rifles and machine-guns fought courageously, being knocked off like nine-pins by the machine-guns of the passing tanks. One by one the tanks negotiated a road-block of concrete cubes then knocked out another 88 mm gun 400 yards away on the right by the roadside, then a self-propelled gun that open fire on them. Suddenly, the tank men saw Americans huddled in the ditch by the roadside and realised they had made contact with the 504th Parachute Regiment. The huge multi-span Nijmegen Bridge and its half-mile long approaches, last but one of the 'Market Garden' bridges, fell intact into Allied hands at 7.15 p.m. September 20th. Arnhem was only eleven miles away.

It is reported that on meeting General Gavin after this success, General Dempsey, commanding the 2nd Army, said:'I am proud to meet the commander of the greatest Division in the world today'.

Further south greatly strengthened German forces, including Meindl's Parachute Corps, were strongly counter-attacking out of the Reichswald; the Americans were holding them but badly needed their Glider Infantry Regiment due on Lift Three. They arrived, together with the remaining Polish Parachute Battalion, into the 82nd Division's area just east of Grave late on the afternoon of the 23 September. Rather than reinforcing Urquhart or drawing German opposition away from him, or strengthening XXX Corps advance, Browning showed that he was more concerned with holding what he had, conscious that the 1st British

1. **The Scheldt-Meuse Canal Bridgehead** where the Third Army Corps started the attack on september 17, 1944 (Operation Garden).
2. **Valkenswaard** (British cemetery, 222 graves).
3. **Nederweert** (British cemetery, 363 graves).
4. **Eindhoven** (Airborne Monument, Liberation Monument, in Woensel British cemetery, 686 graves).
5. **Mierlo** (British cemetery, 665 graves).
6. **Bridge near Son** spans Wilhelmina canal (destroyed by Germans on september 17, 1944, reconstruction completed on september 19, 1944).
7. **Son and Breugel** (Airborne Monument).
8. **Bridge near Best** across the Wilhelmina canal (destroyed by the Germans on september 18, 1944).
9. **Best** (Joe Mann Monument in Natuurtheater).
10. **St.-Oedenrode** (Castle Henkenshage, HQ of U.S. Airborne Division).
11. **Venray** (British cemetery, 692 graves).
12. **Overloon.** Tank battle from september 26 until october 16,1944. (National War and Resistance Museum. British cemetery, 280 graves).
13. **Veghel** (Airborne Monument).
14. **Heeswijk-Dinther** (Castle of Heeswijk, Airborne Chapel).
15. **German attacks** on the narrow corridor where a number of breakthroughs took place around september 20, 1944 which caused serious delays.
16. **Uden** (British cemetery, 703 graves).
17. **Bridge at Grave** over Meuse River (taken by U.S. 82nd Airborne Division on september 17, 1944).
18. **Lock and bridge at Heumen** across Meuse-Waal canal (taken by U.S. 82nd Airborne Division on september 17, 1944).
19. **Bridge at Malden** across Meuse-Waal canal (destroyed by Germans on september 17, 1944).
20. **Bridge at Hatert** across Meuse-Waal canal (destroyed by Germans on september 17, 1944).
21. **Milsbeek** (British cemetery, 210 graves).
22. **Mook** (British cemetery, 322 graves).
23. **Groesbeek.** (Canadian cemetery, 2595 graves).
24. **Nijmegen-Groesbeek.** Hotel Sionshof, where American paratroopers and Second British Army made the link-up (memorial plaque in housefront).
25. **Nijmegen** (the Jonkerbos British cemetery, 1636 graves).
26. **Roadbridge near Nijmegen** across the Waal river (taken by Allies on september 20, 1944). Time-capsule of Operation deposited in Hunner Park, september 1974.
27. **Railway bridge at Nijmegen** across the Waal River (taken on september 20, 1944, subsequently destroyed by the Germans).
28. **Heroic crossing of the Waal river** by U.S. paratroopers using rubber boats. They took the railway bridge as well as the northern approach to the road bridge on september 20, 1944.
29. **Elst** (Allied offensive bogged down around september 24, 1944).
30. **Heteren** (commemorative plaque in town hall to honor U.S. units).
31. **Driel.** (Monument for Polish paratroopers).
32. **Railway bridge across Rhine near Arnhem** (destroyed by Germans on september 17, 1944).
33. **Road bridge across Rhine near Arnhem** (of "A Bridge too Far" fame: the bridge that despite heroic fighting by the 1st British Airborne Division remained in German hands).
34. **Oosterbeek.** (British Airborne Cemetery, 1746 graves).
35. **Oosterbeek.** (Airborne Monument opposite Hotel 'Hartenstein', HQ of Major-General Urquhart during the Battle of Arnhem).
36. **Doorwerth.** (Airborne Museum in Doorwerth Castle).
37. **Heelsum.** (Airborne Monument).
38. **Ede.** (Airborne Memorial and two memorial stones on Ginkelseheide opposite Hotel De Ginkel).
39. **Oosterbeek-Laag.** (Benedendorpse kerk, last rallying point for paratroopers before their retreat on september 25, 1944).
40. Paratroopers retreating across the Rhine River (september 25/26, 1944). Only 2163 of the originally landed 10,005 Red Devils managed to escape, leaving 1,200 dead, and 6,642 missing, wounded or taken prisoner.

By courtesy of the Netherlands National Tourist Office.

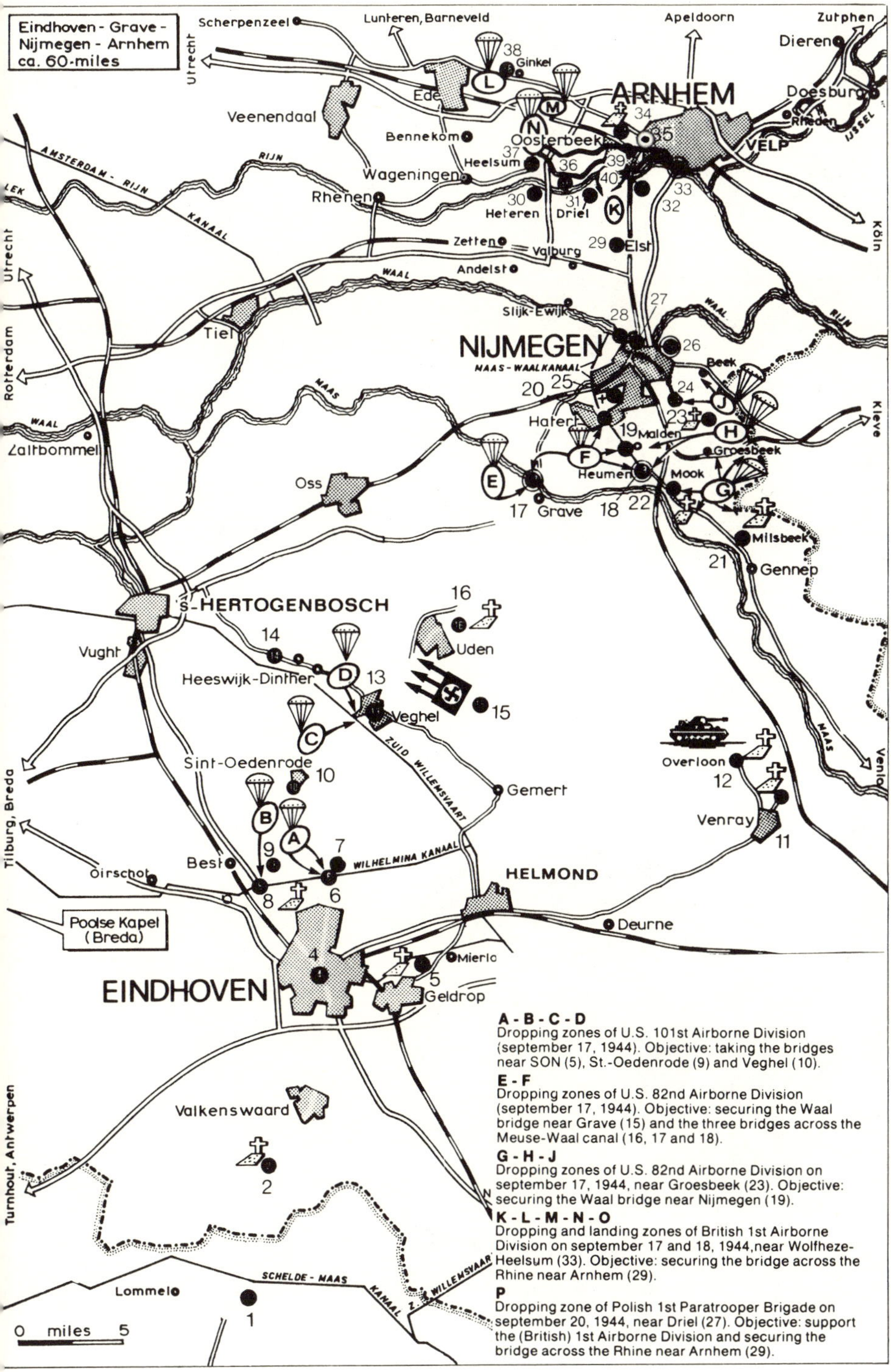
Eindhoven - Grave - Nijmegen - Arnhem ca. 60-miles
Scherpenzeel
Lunteren, Barneveld
Apeldoorn
Zutphen
Dieren
Utrecht
Ginkel
Ede
ARNHEM
Doesburg
Veenendaal
Oosterbeek
Rheden
VELP
IJSSEL
Bennekom
Heelsum
AMSTERDAM - RIJN KANAAL
RIJN
Wageningen
LEK
Rhenen
Heteren
Driel
Köln
Zetten
Valburg
Elst
Andelst
Utrecht
WAAL
Slijk-Ewijk
Tiel
NIJMEGEN
MAAS - WAALKANAAL
Rotterdam
Beek
MAAS
Hatert
Malden
Kleve
WAAL
Zaltbommel
Groesbeek
Heumen
Mook
Oss
Grave
Milsbeek
Gennep
'S-HERTOGENBOSCH
Uden
Vught
Heeswijk-Dinther
Veghel
ZUID WILLEMSVAART
Sint-Oedenrode
Gemert
Overloon
Venray
Venlo
Tilburg, Breda
Best
WILHELMINA KANAAL
Oirschot
HELMOND
Poolse Kapel (Breda)
Deurne
Mierlo
EINDHOVEN
Geldrop
Valkenswaard
Turnhout, Antwerpen
Lommel
SCHELDE - MAAS KANAAL
Z. WILLEMSVAART
0 miles 5
A - B - C - D
Dropping zones of U.S. 101st Airborne Division (september 17, 1944). Objective: taking the bridges near SON (5), St.-Oedenrode (9) and Veghel (10).
E - F
Dropping zones of U.S. 82nd Airborne Division (september 17, 1944). Objective: securing the Waal bridge near Grave (15) and the three bridges across the Meuse-Waal canal (16, 17 and 18).
G - H - J
Dropping zones of U.S. 82nd Airborne Division on september 17, 1944, near Groesbeek (23). Objective: securing the Waal bridge near Nijmegen (19).
K - L - M - N - O
Dropping and landing zones of British 1st Airborne Division on september 17 and 18, 1944, near Wolfheze-Heelsum (33). Objective: securing the bridge across the Rhine near Arnhem (29).
P
Dropping zone of Polish 1st Paratrooper Brigade on september 20, 1944, near Driel (27). Objective: support the (British) 1st Airborne Division and securing the bridge across the Rhine near Arnhem (29).

The Arnhem Wargame. The 17 pdr anti-tank gun defending the western end of the road running past the church in the Oosterbeek perimeter.

The Arnhem Wargame. The 17 pdr anti-tank gun and parachutist heavy machine-gun team behind earthworks in the Oosterbeek perimeter

The Arnhem Wargame. Western corner of the Oosterbeek perimeter, with 17pdr anti-tank gun dug-in on the road and Lonsdale Force defending the churchyard wall.

Airborne Division would have to be evacuated as soon as possible.

At Arnhem, after Frost's resistance ended on the bridge, German General Harzer began flat-out attacks on Urquhart's defences, turning Oosterbeek into blazing rubble as H.E.,phosphorous and mortar shells poured into the area at a rate of more than fifty a minute. Launched simultaneously from two sides of the perimeter, frequent German attacks by tanks and groups of 20/30 infantry were repulsed by Vickers and Bren machine-guns, mortars, rifles, Piats and grenades, while often parachutist sharp-shooters brought German snipers down out of the trees. British anti-tank guns were being knocked out one by one, but Piats fired at point-blank range destroyed numerous German tanks.

The South Staffs withdrawing from Arnhem were rallied by Major Cain and successfully attacked Den Brink, the wooded hill position commanding the Oosterbeek-Arnhem road and then took up positions in front of the white church in Oosterbeek. Almost at once tanks and self-propelled guns began to edge towards their position and 22 year old Lance-Sergeant Baskeyfield won a posthumous Victoria Cross when he destroyed

The Arnhem Wargame. The eastern end of the perimeter at Oosterbeek; the 17pdr anti-tank gun dug-in on the road and Lonsdale Force defending the churchyard wall.

two Tigers and at least one self-propelled gun at 100 yards range with his 6 pdr anti-tank gun, loading and firing single-handed when his crew were cut down by intense close-range fire. The gun was put out of action and the wounded NCO crawled to a nearby abandoned 6 pdr and knocked out another self-propelled gun, before being killed by a tank shell.

On the 21st, the Germans made their first really determined effort to break into the perimeter, being ejected at bayonet point by the 1st Borders. The woods were full of tanks and armoured cars prowling about, and enemy snipers and machine-gunners fired from the trees. The White House Hotel changed hands several times during the day before being finally recaptured at the bayonet point by the KSOB. Casualties were heavy and by the 22nd, 156 Bn. Parachute Regiment was down to 100 men, and the 10th Bn. was 30 men strong with no officers remaining. Also on the 22nd, Major Lonsdale's force began fortifying the church, and their commander preached them a sermon from the pulpit.*

* This was scrawled on the church door and can now be seen in the Airborne Forces Museum at Doorwertsch, near Arnhem.

The Arnhem Wargame. View of the perimeter defenses and Oosterbeek church.

The Arnhem Wargame. The church and road defenses at the southern end of the Oosterbeek perimeter.

'You know as well as I do that there are a lot of bloody Germans coming at us. Well, all we can do is to stay here and hang on in the hope that somebody catches us up. We must fight for our·lives and stick together. We have fought them in North Africa, Sicily, Italy, at times against odds. They were not good enough for us then and they are not bloody well good enough for us now. An hour from now you will take up defensive positions to the north of the road outside. On these positions we must stand or fall and shoot to the last round. Make certain you dig in well and that your weapons and ammo are in good order. We are getting short of ammo so when you shoot, you shoot to kill. Good luck to you all'.

On the afternoon of 21 September, their dropping-zone switched to the south side of the river at the Heveadorp Ferry, the unfortunate Polish Parachute Brigade were again affected by bad weather which caused nearly half their C47s to turn back without dropping their men. General Sosabowski, after landing against light opposition, found himself on the south bank of the river with only 750 men. Assembling his force and moving to the bank, he discovered that the British had been driven off the small hill that dominated the area and had lost the ferry. So, with no boats or rafts to ferry them across, the Poles established a defensive area at Driel.

Now the Germans were able to push forces across the Arnhem road bridge to reinforce the 10th SS Panzer Division who had been ferried across the river earlier. These increased armour and anti-tank forces held up the Guards Armoured Division and Horrocks replaced them with the 43rd (Wessex) Infantry Division; but their training and experience caused the infantry to remain on the roads and to expect tank and artillery fire support so that they advanced slowly, and it was nearly dark before their leading battalion reached Driel and contacted Sosabowski's Poles. Early on the following day, the 22 September, reconnaissance elements of the Guards Armoured Division, advancing out of the Nijmegen bridgehead, contacted the Poles. Subsequently General Horrocks conferred with Sosabowski, Browning and General Thomas of the 43rd Wessex Division, when it was decided to send the Poles across the river, although all hopes of establishing a northern bridgehead were abandoned.

The Arnhem Wargame. Oosterbeek church defended by Lonsdale Force, with the Arnhem perimeter stretching to the far horizon. See also photograph on page 188.

The Arnhem Wargame. Lonsdale Force defend the churchyard at the southern end of the Oosterbeek perimeter.

Each day, when weather permitted, R.A.F. Transport aircraft courageously flew in and dropped supplies to Urquhart's force, suffering 20% losses in aircraft from AA fire and German fighters. Tragically, only 7% of these supplies landed within reach of the garrison, because the perimeter was continually shortening and it was impossible to re-direct the pilots.

The 23rd saw the heaviest German onslaughts fall on Lonsdale's sector, where Major Cain of the South Staffs won the Victoria Cross and lived to tell the tale. First he immobilised a Tiger tank with a Piat fired from the Red House at 20 yards range; then, although wounded by machine-gun fire and collapsing brickwork, he crept out and brought up a 75 mm gun that completely destroyed the Tiger. Later he drove off three tanks with a Piat, and on the 25th manned a 2 inch mortar and repelled an attack by SPs and infantry with flame-throwers.

The beleaguered Airborne troops were now receiving fire-support from Spitfires and Typhoons, escorting supply aircraft and strafing German gun positions. A4·5 inch artillery battery south of the river, directed by Airborne OPs in the perimeter, started directing a steady stream of shells at enemy targets; then the gunners erected a 25 foot wireless aerial that gave better reception to the radio messages coming from the perimeter, and increased their fire with two batteries of 5·5 inch and 155 mm guns.

But the impetus of XXX Corps attack had petered out and heavy flank attacks had cut their slender supply line in several places. By now, even if the Arnhem road bridge had been secured, it is unlikely that British 2nd Army would have been able to exploit the situation.

On the night of 22nd, General Urquhart had sent officers across the lower Rhine in inflatable dinghies to report the worsen- situation to General Horrocks. They returned on the following night and confirmed that the 2nd Army were remaining on the south bank, but that General Dempsey wished the perimeter to be reinforced before attempting a withdrawal. A small group of the 3rd Polish Parachute Bn. had already managed to cross the river in boats and on rafts and, at midnight on the 23rd, some more Poles and 5th Bn. Dorset Regiment attempted to cross near the Heveadorp ferry. Detecting them, the Germans raked the boats with machine-gun fire so that only 250 Poles and 350 Dorsets managed to get across. By this time, 1st Airborne Division was down to about 2,500 men, short of ammunition and supplies – and very tired.

Urquhart began planning to evacuate the perimeter, but the actual decision to withdraw was taken on 25th September by General Browning on General Dempsey's authority. Considerable fighting continued throughout the 24th and well into the 25th when at 1830 hours Urquhart's officers were told to prepare to move out. The northern positions, some two miles from the river, were to be evacuated first, gradually progressing southwards so that the last to leave would be the men nearest the riverbanks. All the doctors volunteered to stay with the wounded, who had to be left behind.

At 2200 hours the withdrawal began along the designated escape corridor, with glider pilots posted as guides. Faces were blackened, boots muffled and loose equipment tied tightly as the first men left their posts. In pitch-blackness and pouring rain men moved in single file, each man grasping the smock of the man in front, shuffling along through the muddy Dutch ploder-land down to the river banks, to lie in wet slime waiting their turn, with mortar bombs, shells and machine-gun fire raining down upon them. Earlier, there had been only desultory German shelling and mortar fire, and British gunners on the south bank fired on enemy positions flanking the perimeter to keep them quiet.

Inevitably, there were not enough boats, and some were sunk as they plied back and forth, so that when dawn came several hundred men remained unevacuated; some attempted to swim the river and were drowned. Realising what had happened, Germans swarmed into the abandoned perimeter and the early morning air resounded to the crash of tank guns and small-arms fire as isolated parties of trapped defenders made their last stands. Then it petered out and for the first time in eight days, Arnhem was quiet.

Why had such an adventurous and promising enterprise failed? There are many reasons that singly were serious and collectively fatal. The advance of XXX Corps, acknowledgely

The Arnhem Wargame. Major Lonsdale's Airborne troops defend Oosterbeek church at the Rhine end of the perimeter.

difficult was not as rapid as it could have been, particularly as 82nd and 101 U.S. Airborne Divisions had provided a continuous carpet almost as far as Nijmegen. If Horrocks force had advanced as rapidly as predicted then Frost's small force on the bridge might have held long enough to complete the link up. Or, if all 1st Airborne Division had reached the bridge as planned, there is little doubt that they would have held out until XXX Corps arrived, even though the delay at Nijmegen caused them to take seven days to reach the Rhine.

The operation was gravely handicapped by decisions over Lifts Two and Three having to be taken without proper knowledge of the tactical requirements on the ground.

The airborne technique in 'Market Garden' was good, units and formations being dropped or landed with unparalleled precision and concentration, indicating that there was excellent co-operation between airborne units and troop-carrier squadrons. But it can be claimed that the Airborne forces had some cause to complain that offensive air-support – their 'sword and shield' – was poorly provided.

The operation involved many notable ground actions that will go down in military history, with that of the 82nd U.S. Airborne Division being almost classical. In the south, the airborne operations were a complete success; in the centre they did not fully succeed but were still first-class; in the north, in the strict sense of the plan, they just about succeeded through the courageous efforts of Frost's 2nd Parachute Bn. Operation 'Market Garden' was a failure in a number of tactical respects but, rather than battlefield errors, its strategic failure can be blamed more on its timing, and the energetic recovery of an underestimated enemy.

AIRBORNE OPERATIONS RECONSTRUCTED AS WARGAMES

The three components involved in airborne warfare are – 1. paratroopers; 2. glider-borne troops, and 3. carrier, tug or transport aircraft – each or both of the first two, in association with the third, are essential factors in this type of warfare when simulated on the wargames table. Without the aircraft simulated airborne operations, although a different and specific type of wargame, on the table-top, become conventional wargames involving lightly-armed elite troops, initially blessed by surprise, against increasing numbers of conventionally-armed troops supported by armour, artillery and often air-power. Whilst acknowledging that these factors might make for a different style of wargaming, the full flavour of the more intriguing aspects of airborne warfare is only savoured when actively prefaced by a simulation of para-drops and glider landings. Although a war-game narrative can either begin with airborne forces about to drop or else with them already in position – in both cases it is possible to authentically reconstruct actual historical operations such as Crete or Arnhem.

Scaled-down numbers

Table-top reconstructions may take the form of 'one for one' with the same numbers of figures on the wargames table as took part in the original action so that, if re-fighting Bruneval, 'C' Coy. 2 Para. Regt. would consist of 130 figures divided into 'Nelson', 'Drake' and 'Rodney' groups as in real life. Or with one figure on the table representing say 20 actual men – thus, 215 figures take the part of the 4,300 German paratroopers who fought at Crete for the first eight hours. Both methods affect the load capacity of the aircraft engaged – the twelve Whitley carrier-planes that took John Frost's raiders to the French coast can be reproduced over the wargames table by twelve 'simulated' Whitleys. But the Germans used ten troop-carrier groups each of 50 Ju52's and DFS230 gliders – an aircraft lifted 13 paratroopers, the glider carried a pilot, co-pilot and nine men. So, scaled-down to 1 = 20, each Ju52 will carry 13/20th's of a man and each glider just over half! Obviously the answer is to scale down the number of aircraft and let each carry its specified real life load.

Using a points system

A points system that allows a glider to carry a suggested 20 points has values allocated as follows:-

An airborne soldier = 1 point
Ammunition, fuel and ration containers = 2 points
6 pdr Anti-tank gun = 6 points
17 pdr Anti-tank gun = 10 points
Jeep = 8 points
Trailer = 4 points
Mortar = 4 points etc., etc.

Under these circumstances, the wargamer has to decide how best to utilise the capacity of his aircraft and gliders to land on the ground the most favourable combination of men and weapons – not an easy decision for real-life generals as the events at Arnhem showed!

When considering a table-top airborne operation, the 'commander' of the raiders assesses the available aircraft and their loads, and decides whether there is to be more than one lift (there has to be half a day, in wargames-moves 'time', between them). With points-values given to EVERYTHING – aircraft, gliders, men, weapons and equipment – a 'commander' can choose his requirements up to his allocated points total. Order of Battle and Objectives settled, dropping and landing zones scaled-down to 200 ×200 mm (8 × 8 ins.) areas are selected by allowing a brief glance at the table-top terrain, as though glancing at a map.

Zone marked by coloured-head map pins stuck into terrain.

Simulating the marking of zones

The first operational move is to mark the zones by dropping Pathfinders or beacons, represented by artefacts, then the areas are marked out by coloured-head map-pins stuck into the table-top. The counters representing the Pathfinders are of a heavier nature so that they fall accurately instead of drifting over the terrain. The Pathfinders are allowed a set number of game-moves (say three) to set up their beacons and this opening gambit when opposed provides a minor curtain-raiser to the action. The flare-beacons favoured by the Germans can be simulated by dropping weighted markers that mark the zones by remaining where they fall. An interesting side issue can be the presence of Resistance-fighters, Partisans or Guerillas who douse the beacons as soon as they fall!

Pathfinders failing to lay markers, or beacons not correctly dropped result in confusion simulated by the wargamer carrying out the dropping procedure in the dark! After a brief glance at the terrain (to represent map study) the wargamer positioned by the wargames table, carries out the chosen dropping method with the lights switched off.

German flare-beacons simulated by suitably coloured plastic markers.

The area of wargaming operations

The wargame may require a map or maps of the area of operations made up of a number of wargames tables; the actual wargaming might be taking place on table 3, while off-target landings are in adjacent tables nos. 2 and 4, from where paratroopers make their way at an agreed rate of map movement towards the operational wargames table. This particularly applies when reconstructing large airborne operations such as 'Market Garden' at Eindhoven, Niejmegen and Arnhem, where the entire area is far too extensive to be laid out on a single table-top terrain. An interesting and realistic wargaming simulation of an airborne operation accepts the existing wargames table as the entire area of operations, with all landings taking place within its bounds. This is done by simulating drops so that paratroopers are scattered over the terrain, or by assuming that the drops have already taken place and that the paratroopers are in position or moving off to their objectives.

The time-chart

A time-chart is required, detailing the numbers of game moves for each phase of the operation, which have to be carried out regardless of whether preceding scheduled tasks are completed or not.

A Dakota, made-up from a plastic kit, used to drop paper parachute-shapes simulating a para-drop.

Simulating para-drops

Now comes perhaps the most important single item in wargaming airborne operations – simulating the actual para-drop and glider landings. The simplest method is to tacitly assume that they have already taken place and the paratroopers are on the ground, although distributed in a random manner as in real warfare. The dropping zone decided, a dice score of 4, 5 and 6 for each man or group means that they have landed on the D.Z.3 = 30 cm (12 ins), 2 = 45 cm (18 ins) and 1 = 60 cm (24 ins) from the centre of the zone – in the direction the wind is blowing. Or use percentage-dice – 50% and below indicating a landing in the zone, with 15 cm (6 in) drift for every 10% above 50.

The model soldiers themselves are not dropped onto the wargames table, each man or group being represented by a paper 'parachute-shape' or a plastic counter.

Time and convenience preclude bringing over the table ALL the aircraft, even though their numbers have been drastically scaled-down. Representing the simultaneous dropping of the entire force, ALL 'parachute-shapes' or counters carried in a single 'aircraft' representing every 'plane that should be over the table at one time. Paper calculations determine aircraft losses during the run in, stricken craft being deleted from the total availability of aircraft and an equivalent number of counters removed from those placed within the 'aircraft'.

A visually attractive simulation of an air drop is to use a 1:72 scale plastic kit of a C47 or Ju52, with a 2 cm (¾ in) square opening cut in the fuselage belly for loading the counters, and for ejecting the 'paratroopers' by tilting the nose downwards as the model moves across the table. It is held in the hand at a steady 60 cms (24 ins) above the terrain, representing 100 feet at 15 cms (6 ins) = 400 feet. A length of string threaded through a hook screwed into the top of the fuselage can have one end fixed 30 cms (12 ins) above table level, and the other end held in the air so that the weighted model slides gracefully across the table, its contents ejecting by their own weight as the nose dips and the string is jerked.

Although requiring a certain lowering of standards of realism, it is far simpler and perhaps more effective to replace the model aircraft with a lidless cardbox containing 'parachute-shapes' – paper circles about the size of a 1p piece cut from flimsy paper so that they 'flutter' separately down to earth rather than dropping in a bundle. At the scaled height representing 400 feet the wargamer holds the box/aircraft over the area where he wishes to drop his paratroopers. Then turning the box upside-down sends an entire airborne brigade or division fluttering down to lie scattered realistically on and around the dropping-zone.

The flimsy nature of the shapes causes them to disperse over an area that widens in direct relationship to the height from which they are dropped – the lower the box is held the more compact the drop and, conversely the shapes become more scattered if the box is held higher. However, a parachute does not have time to open if the drop is too low, so 25%drop casualties are incurred on drops between 400 and 300 feet, 50% casualties at 300-200 feet and total loss below that height. The 'Armour and Infantry 1925-1950' Rules of the Wargames Research Group include a chart controlling the landing of men and equipment in the open, in steep slopes or among rocks, in woods or among buildings, water or swamp – graduated totals having to be attained by dice-throw to determine whether the man or equipment is safe.

Variations in height over the dropping-zone are decided by Chance Cards: a set of ten cards, six marked 400 feet; two indicating that the aircraft has been forced higher by anti-aircraft fire and will be dropping its cargo at 500 feet; and two indicating the presence of enemy fighter aircraft that force the carrier-aircraft to come in at low altitude – a dice-throw decides that height i.e. 4, 5, 6 = 350 feet, 3 or 2 = 250 feet; 1 means that the nervous pilot has ejected his paratroopers at under 200 feet. An aircraft drawing any score but the fatal 1 can opt to 'go round again' and try again to come in at 400 feet. This means that it will again have to run the gauntlet of ground anti-aircraft fire for as many moves as it takes to cross the area.

By being coloured differently, counters or 'parachute-shapes' indicate what they represent: white is a paratrooper armed with

Cardboard box containing parachute-shapes held over terrain prior to being tilted to release parachutists.

Parachute shapes fluttering to ground from tilted box.

Parachute shapes dispersed on terrain as they fell.

Parachute shapes replaced by groups of 25mm parachutist figures.

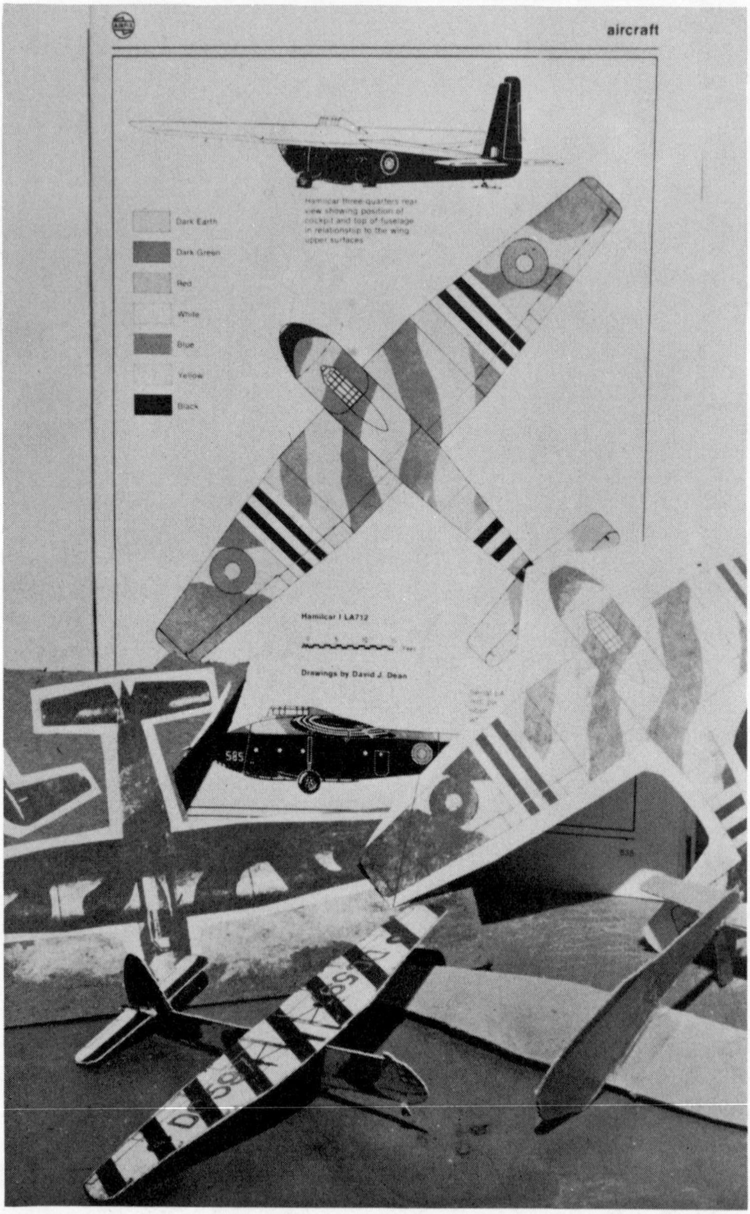

Steps in constructing a simple glider, beginning with photo-copying line-drawings from a magazine; sticking onto card; cutting-out and fixing wings and tail to fuselage with cellotape. All the gliders shown in the photographs of this book were constructed in this simple fashion.

Allied gliders and airborne troops in 1:300 scale (supplied by Heroics and Ros Figures of London).

the usual assortment of weapons; red an officer; yellow – mortar crewmen; light blue – machine-gunners; green – anti-tank projectile men (bazooka, PIAT or panzerfaust); dark-blue – flamethrower operators; and black is the container that has to be reached by the last four categories before they are armed.

Strong winds were dreaded and played havoc with airborne operations as in the Sicily landings. When wargaming it has to be decided whether wind is to be considered broadly or locally, the former applies to campaigns using maps when wind can divert a drop many miles from its intended dropping zone; local effect is solely on the single table-top terrain causing paratroopers to be scattered. In either case, first orientate the map or table-top terrain, then draw from marked cards to decide the direction the wind is blowing. Next dice – score 4,5 or 6 = no appreciable wind; 3 = a light wind; 2 = a stronger wind; and 1 = gale-force wind. No wind allows paras or gliders to land in designated zone; light wind – they are dispersed 10 cm (4 ins); stronger wind disperses them 20 cm (8 ins); gale-force winds cause 50% casualties. All dispersal movements are in the direction the wind is blowing. In a 'local' context, the flimsy shapes inevitably scatter so such dispersal activities may not be considered necessary.*

Simulating Glider Landings

The variation in cargoes carried by gliders precludes having a single model represent ALL aircraft. Instead scaled-down numbers of gliders are employed, each with details of its cargo of infantry, artillery, vehicles, etc.

The simplest means of simulation is a matchbox (or similar container) holding counters inscribed 'rifleman' ; 'jeep' ; '6 pdr anti-tank gun' ; '17 pdr anti-tank gun' ; 'field gun' ; 'pilot' ; 'co-pilot' ; etc., etc. The box is dropped onto the table or, if some skill is required, the wargamer stands clear of the side of the table and lightly tosses the box onto the table in an effort to land it as close to its designated zone as possible. In any event, its resting place can be subject to the same sort of vagaries (i.e. Chance-cards and varying strength winds) that affect carrier-aircraft and paratroopers.

The same conscience that causes wargamers to go to extreme lengths in producing authentically dressed and equipped armies will almost certainly result in practical attempts to provide gliders that *look* realistic, even though not actually *carrying* their specified cargoes. On the table-top space precludes the use of too many gliders of 1:72nd scale, models should be that much smaller in size as to be almost symbolic. Line-drawings and photographs of gliders can be traced or photo-copied, stuck onto card and then made up into single dimensional models, or with wings set upon fuselage. On a blank reverse side can be lightly pencilled details of their contents. Model gliders can be turned out by mass-production methods using card or plastikard templates from which three or four at a time can be cut out of thin card. Thus the glider availability of a force can be dramatically increased in a few minutes!

*The simulation of general weather conditions, including winds, in relation to the seasons of the year is described in considerable detail in the books *Air Wargames* – pages 144-148, and *Advanced Wargames* – pages 197-204, by Donald Featherstone; published by Stanley Paul of London.

Gliders with single card thickness fuselage and wings can be given substance by backing the photo-copied or drawn exterior with polystyrene, rounded off with sandpaper to give the impression of bulk. A small square of plastikard let into the under surface records details of the glider's contents, erased after the operation.

Gliders are assumed to be released from their tug-aircraft 'off the table', coming in to land on a single game-move. Make a 'Landing zone chart' scaled so that each of its nine numbered squares equals 10 × 10 cm (4 × 4 ins) on the actual table, with centre square No. 5 representing the specified landing zone and left clear on the table, whereas other squares can include obstacles. When a glider comes in, a percentage dice indicates the numbered square in which it rolls to earth ; should it strike an obstacle in landing, an ordinary dice indicates its fate : 1 = total destruction of glider and contents ; 2 = too badly damaged to allow heavy weapons or vehicles to be unloaded ; 3 = some of its human load are injured and must be diced for — 4, 5 or 6 saves them (do not forget to include the pilot and co-pilot as they fight once on the ground!) A glider landing in the sea, in a lake or river means all equipment is lost and every man on board has to score 4, 5 or 6 on the dice to save himself.

A modicum of skill is required to represent the glider by a folded paper aeroplane (marked with its contents) deftly thrown onto the table – this may well appeal to the schoolboy wargamer who regularly practises in the classroom!

Effects of anti-aircraft fire

Enemy anti-aircraft weapons may be assumed to be 'off-table', 'paper' guns of predetermined number, their effects calculated on paper – each machinegun is worth 1 point and each anti-aircraft gun 3 points. With all carrier-aircraft and gliders over the table within range, throw a dice for each weapon – a hit is represented by each 6 for an A/A gun (scoring 3 damage-points) and each 5 or 6 for a machine-gun (scoring 1 point). An aircraft or glider is worth 10 'damage-points', and is destroyed when that total is reached ; dice being thrown for occupants – 4, 5 or 6 indicates that paratroopers have baled out of a stricken carrier, but glider-borne troops require 6 to land safely ; all equipment is destroyed.

Simulation of bombing and air-strikes

This aspect of air warfare requires attention because it was the 'artillery of the airborne forces' ; also, air-attacks were frequently made on paratroopers fighting on the ground. The simplest method of simulating air-strikes (described at far greater length in the book *Tank Battles in Miniature: The Western Desert Campaign* by Donald Featherstone (Patrick Stephens Ltd. 1973) is for attacker and defender each to choose a number from 1 to 10 – if both choose the same number, the attacking aircraft is shot down ; otherwise it inflicts casualties to an agreed scale on its ground target. Bombing is simulated by using a bomb-carpet template – a 25 cm (6½ in) square of clear plastic, perforated by nine 2 cm (¾ in) circles numbered 1–9 – the centre of the template is placed on the point where bombs are to be dropped and a percentage dice is thrown. The bomb drops in the circle bearing the number thrown on the dice and anything covered or partially covered by the circle is hit. Bombers can be damaged in the same way as carrier-aircraft and gliders.

The Landing

The actual drop counts as 1 game-move when defenders can fire on descending men, 1 dice per gun – 6 to hit. At this stage casualty calculations are made with pencil and paper, losses deducted when counters are replaced by actual figures on the terrain. There were occasions when paratroopers fired their personal weapons whilst dropping and they may do so after fire has been taken from ground defenders ; the likelihood of their random fire scoring a hit should be reflected in the rules.

Hitting the ground, paratroopers take a round of fire from defenders while disentangling themselves from their parachutes and seeking containers of heavy weapons. They may either move or return fire on an individual basis, fighting as individuals ; they are not organised until the move after landing. Moving at 1½ times normal movement rate, they seek containers, open them, arm themselves and are ready to fire by the end of the move.

Men who drop into sea, lakes, rivers or swamps are saved by dice throw of 5 or 6 – in the first two situations providing a boat is nearby to pick them up ; they come out of river or swamp on the side to which they are nearest. Dropping into a minefield detonates a mine which kills the man ; entangled in the higher branches of a tree or on a roof requires 4, 5 or 6 to cut clear and climb-down, taking two moves.

Ground fighting

Airborne troops were selected men, highly trained and battle-hardened, so must classify as ELITE forces on the wargames table, allowed bonuses in move-distance (representing speed of movement), and in firing and fighting capabilities. Encountering enemy troops, airborne troops fire first, except when enfilade (flanking) fire is directed upon them.

Being the most readily obtainable figures, the AIRFIX range of Allied and Axis paratroopers will figure prominently in wargames reconstructions of airborne operations ; each of these individual figures is armed with a specific weapon such as sub-machinegun, rifle, grenade, etc. – it should not be mandatory for him to use only that weapon. Rather the wargamer should have the choice of using any of the weapons he carried, so that a figure in an action-pose of throwing a hand-grenade might actually throw it on move 1, fire a sub-machinegun on move 2 and use a knife in a melee in move 3. The same facilities may apply to opponents, as it exemplifies the diversity of the modern infantryman's arsenal.

Movement	*Movement per game-move*
Airborne soldier – paratrooper, gliderborne infantry	180 mm (7 ins)
Airborne mortar or heavy machinegun team	150 mm (6 ins)
Infantryman	150 mm (6 ins)
Infantry mortar or heavy machinegun team	120 mm (4½ ins)
Jeep, or similar towing vehicle	450 mm (18 ins)
Motor-cycle	450 mm (18 ins)
Bren-carrier	450 mm (18 ins)
½-track vehicle	350 mm (14 ins)
Armoured car	350 mm (14 ins)
Light Tank	350 mm (14 ins)

Medium Tank	300 mm (12 ins)
Heavy Tank	250 mm (9 ins)
Self-propelled gun	300 mm (12 ins)
88 mm gun and tow-vehicle	300 mm (12 ins)

Firing – Small Arms	*Range*
Pistol	75 mm (3 ins)
Rifle	450 mm (18 ins)
Sub-machinegun	100 mm (4 ins)
Light machinegun	300 mm (12 ins)
Heavy machinegun	500 mm (20 ins)
Hand Grenade	75 mm (3 ins)
Flamethrower	75 mm (3 ins)
Panzerfaust, bazooka, PIAT	75 mm (3 ins)

Whenever possible, fire groups of men rather than individuals, using the procedure for a single man firing multiplied by the number in the group. Thus, a rifleman or group nominate the target and throw 1 percentage-dice, from its score is deducted –

1 – if the target-figure is moving
1 – if the target-figure is behind hard-cover (walls, in houses, etc.)
1 – if the target-figure is lying-down.

If firing at close range, add 1 to dice-score. To hit the target a total score of 6 or over is required. A light machinegun counts as three men ; a heavy machinegun as six ; sub-machinegun as two.
Hand Grenade – use burst-pattern as for artillery-firing – circles 15 mm (¾ in) in diameter and 1 cm (¼ in) apart.
Flamethrower – cut a flame-pattern from card or plastic, triangular in shape with sides 7·5 cms (3 ins) long and base 2 cms (¾ in) ; it can be suitably embellished with red and black colouring. Placing point against nozzle of flamethrower, lay the pattern out in front of figure, everything touched by pattern is killed and all men within a move-distance MUST withdraw half-move distance from present position. Carriers and soft-vehicles touched by flame automatically burn ; armoured vehicles do not burn but must move out of range. Troops in houses under flames must

evacuate within a game-move if flaming continues.

Firing – Artillery and Mortars

All artillery (including mortars) can reach *any* point on the war-games table. Firing is either 'open-sight' or 'observed' – the former when the target is visible to the gunners from the site of the gun or mortar ; the latter when the target is not visible to the gunners (behind a hill, etc.,) so that a radio-equipped observer has to be positioned (preferably concealed) with a range of vision that enables him to direct his gun or guns onto the target.

Both types of firing are simulated in much the same way – Burst Patterns. Squares of clear plastic perforated by a pattern of eight numbered circles whose diameter varies according to the weight of the missile are laid with the centre circle covering the designated target ; a percentage-dice is thrown and the missile lands in the circle whose number corresponds with that shown on the dice. Everything within the circle or touched by its rim is considered hit. OBSERVED fire first has to range on target, requiring 5 or 6 on an ordinary dice ; once ranged, gun does not have to range again on same target unless the gun moves. When ranging, dice scores under 5 or 6 indicate a miss, and centre-circle of burst pattern is placed over points as follows: dice score 1 = 15 cms (6 ins) in front of target ; 2 = 15 cms (6 ins) behind target ; 3 = 15 cms (6 ins) to right of target ; 4 = 15 cms (6 ins) to left of target. If observer is killed, then a fresh observer has to reach viewing-position before guns can again be ranged. Arranged in three lines of three, the burst circles are 2 cms (¾ in) apart (from circle-rims) and are in the following sizes to represent different missiles –

50 mm (2 in) Mortar	2 cms (¾ in)
75 mm Recoilless Gun	3 cms (1 in)
75 mm Pack Howitzer	3 cms (1 in)
81·4 cm (3 in) Mortar	3 cms (1 in)
105 mm (4·2 in) Mortar	4 cms (1½ ins)
105 mm Recoilless Gun	4 cms (1½ ins)
105 mm Howitzer	4 cms (1½ ins)

Firing – Anti-Tank Guns

This will principally concern the firing of airborne anti-tank guns on whatever armour was brought against landed airborne troops (at Arnhem it included German Tiger tanks) because only rarely was the Tetrarch tank brought in by glider and its thin armour could be penetrated by even the 20 mm gun at 1300 yards. Against armour, Allied airborne troops could muster 37 mm (2 pdr) ; 57 mm (6 pdr) ; 76·2 mm (17 pdr) and the infantry anti-tank projectors, bazooka and PIAT. Accepting that every gun might be required to fight-off armour, the Germans equipped them all with some sort of anti-tank projectile. They used the 28 mm Schwere Panzerbuchse 41 ; 37 mm Pak 36 ; 50 mm Pak 38 ; 75 mm 40/L/10 recoilless ; and had their airborne programme continued, undoubtedly 75 mm Pak 40 and 88 mm guns would have been brought in. German infantry were equipped with the Panzerfaust anti-tank hollow-charge projector.

Listing the armour-piercing capabilities of the various anti-tank guns is made complex by such factors as the calibre and missile-weight of the firing gun, the angle of impact, the angle of the armour-plates where the missile strikes, the thickness of armour on various parts of tank hull and turret, and the distance between the gun and its target. For practical wargaming purposes, methods of realistic simulation are described in readily available books at far greater length than is possible in these pages.* A reasonable method of simulation can be based on percentages – in World War Two it was known that anti-tank guns had first-time chances of hitting –

At 500 yards (500 mm – 20 ins – on the wargames table)
70% chance, at 2nd attempt 95%
Between 500/1000 yards – 32% chance, at 2nd attempt 90%
Over 1,000 yards – 10% chance, at 2nd attempt 80%

This is simulated on the table-top by throwing a pair of percentage-dice and scoring a hit if they fall at or over the percen-

*For example, see Airfix Magazine Guide 15 *World War Two Wargaming* by Bruce Quarrie, or *Tank Battles in Miniature – The Western Desert Campaign* by Donald Featherstone – both published by Patrick Stephens Ltd.

tages shown above – extraneous factors being represented by percentage deduction viz.

(a). Firer (tank or S.P. gun) is moving at time of firing – 20%
(b). Target-tank is hull-down (i.e. so positioned that only its turret is a visible target) – 30%
(c). Target-tank is moving – 10%

Also required to be taken into consideration is the size of the firing-gun, and whether the target-tank is light, medium or heavy. Broadly speaking, the principal airborne anti-tank guns could knock out listed tanks at specified ranges.

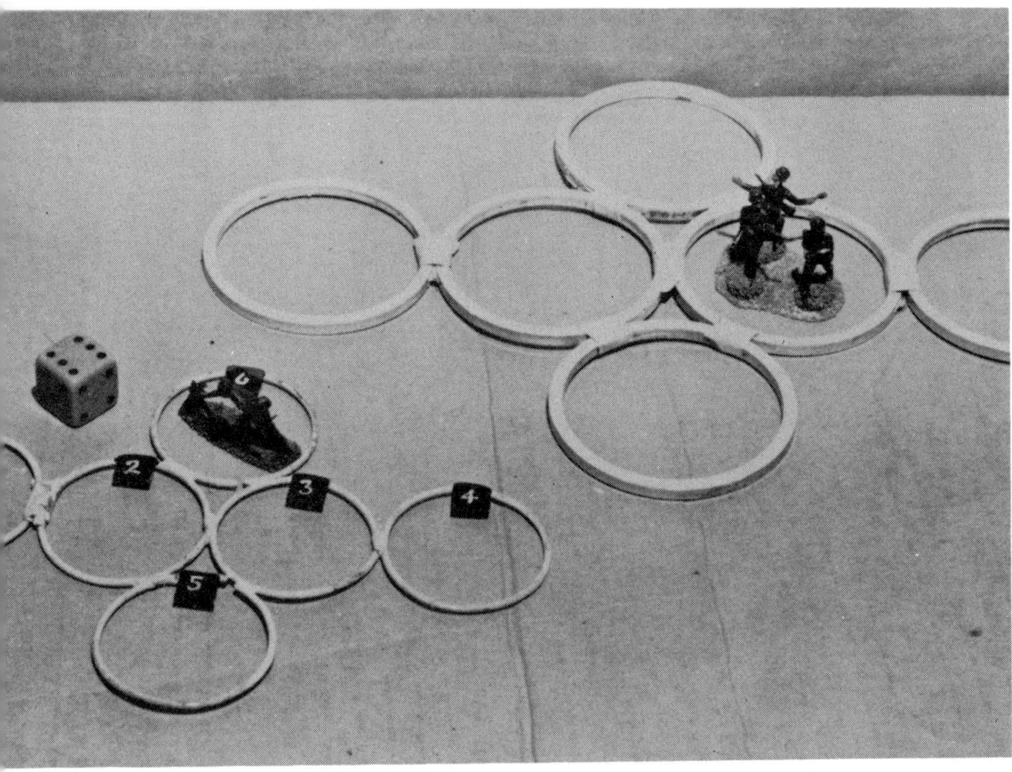

Burst Patterns for mortar and heavy artillery pieces.

	250 mm & under	500 mm	750 mm	1,000 mm	1,500 mm
28 mm Pzbuchse 41	A.10 A.13 Crusader Valentine Tetrarch Honey (Stuart) All Armd/Cars and S.P. Guns	A.10 A.13 Crusader Tetrarch All A/Cs and S.P. Guns			
37 mm Pak 36		As for preceding gun :			
50 mm Pak 38	Matilda A.10 A.13 Crusader Valentine Tetrarch Honey Grant Sherman Stuart Lee All A/Cars All SP Guns	A.10 A.13 Crusader Valentine Tetrarch Honey Stuart All A/Cars All SP Guns	A.10 A.13 Crusader Tetrarch Honey Stuart All A/Cs All Sp Guns	A.10 A.13 Tetrarch All A/Cs All S.P. Guns	
75 mm 40/L10 Recoilless	A.10 A.13 Crusader Tetrarch Honey Grant Stuart Lee Sherman All A/Cars All SP Guns	A.10 A.13 Crusader Tetrarch Honey Stuart All A/Cs All SPs	A.10 A.13 Tetrarch All A/Cs		
75 mm Pak 40	Any Allied tank, A/Car SP Gun	Any Allied tank, A/Car SP gun	All except Cromwell Churchill		

	250 mm & under	500 mm	750 mm	1,000 mm	1,500 mm
88 mm Pak 43	Every Allied tank				All except Cromwell Churchill
37 mm (2 pdr)	Pzkpfw I PzKpfw III PzKpfw IV Marder Stug G III Wespe Hummel Nashorn All/A/Cars	PzKpfw I Marder Wespe Stug G III Hummel Nashorn All A/Cars			
57 mm (6 pdr)	PzKpfw I PzKpfw III PzKpfw IV Marder Nashorn Stug G III Wespe Hetzer Hummel All A/Cars	PzKpfw I PzKpfw III PzKpfw IV Marder Stug G III Wespe Hummel Nashorn All A/Cars	PzKpfw I PzKpfw III Marder Wespe Hummel Nashorn All A/Cars		PzKpfw I Marder Wespe Hummel Nashorn All A/Cars
75 mm M2 & M3	Every German tank *except* Tiger II, Jagdtiger and Elefant	*Except* Panther Tiger I & II Elefant and Jagdtiger		*Except* PzKpfw IVH Panther Tiger I & II Jagdpanther Elefant & Jagdtiger	PzKpfw I Marder Wespe Hummel Nashorn All A/Cars
76·2 mm (17 pdr)	All German tanks	All except Tiger II and		All except Tigers I & II, Elefant Jagdtiger	PzKpfw I PzKpfw III Marder Stug G III Wespe Hetzer Hummel Nashorn All A/Cars

Infantry Anti-Tank Projectors

With a range of only 75 mm (3 ins) on the wargames table, anti-tank projectors must be well concealed to have any chance of success, with armour usually coming to them rather than the infantry stalking the tanks. Any armoured fighting vehicle coming within range of a projector-armed infantryman whether hidden or bravely in the open, takes a big risk of being knocked-out –

At 75 mm (3 ins) range – 75% chance of scoring a hit using percentage-dice
At 50 mm (2 ins) range – 85% chance of scoring a hit using percentage-dice
At 25 mm (1 in) – 95% chance of scoring a hit using percentage-dice

A hit automatically destroys the armoured vehicle, regardless of its size or armour.

In close-country or amid houses, tanks are notoriously 'blind', even so the hidden infantryman must be given a high-degree of concealment-protection under rubble, lichen-moss or other scenic features. Or, balancing the unfair advantage of the wargamer's range of vision as he towers over the table, even if the infantryman is discovered, the tank *must* continue on its intended course UNLESS over 50% is scored on percentage-dice. Wargaming with trusted friends, the hidden projector-man need not be actually placed on the table, but marked on a map or diagram and revealed when action is imminent.

Hand-to-hand Fighting

Generally speaking, in modern warfare there is relatively little hand-to-hand fighting as sheer weight of fire usually forced men to break and run or hug the ground before they could get to grips. This can be simulated by allowing individuals or groups to go-to-ground when coming under fire, then –

1. Once down, they take no casualties but cannot return fire.
2. Still down, they can return fire but take half casualties if fired upon.

OR

3. They remain standing and return fire, causing full-rate casualties to the enemy ; OR move forward their normal distance and fire, causing half-casualties ; OR charge forward with a move-bonus of 50%.

That applies to airborne troops, but line infantry MUST go to ground when coming under fire – they may follow course two if throwing 2 on an ordinary dice and course three if throwing 3, after deducting from the dice-score 1 for any casualties sustained that move ; 1 for coming under fire from approximately twice their own strength, and 1 if under artillery or machine-gun fire.

The unlikelihood of actually coming to grips is simulated by each side throwing an ordinary dice when there is only a normal game-move distance between them, multiplying the number of men in each group by the dice score – airborne troops count 1½ and ordinary infantry 1 point ; troops defending hard-cover (house, wall, etc.,) counting double points value. The group with the lowest total withdraw a move-distance, subject when necessary to the conditions controlling men coming under fire.

Morale

In a sense, these systems replace MORALE – a major factor in warfare and one of the principal means of inducing the ebb-and-flow of combat on the wargames table, where event-controlled scoring methods can force a stricken unit to withdraw or break and run. It is recommended that NO such morale system should be applied to airborne troops for the simple reason that it would be unrealistic ; there is *no* instance in this book of faltering morale on the part of any airborne troops ! Ordinary infantry can be subject to a basic morale check after taking casualties – a suggested method being a dice score of 4 or above, deducting 1 for each THIRD of the group who are casualties. A score of 2 or 3 causes the group to withdraw their normal move-distance in good order; a score of 1 or 2 means that they turn and run a move.

Prisoners of War

Their morale never in question, airborne troops sensed when resistance was useless and surrender offered the chance of living to fight another day with escape possible. Such were the reasons for submissions like those of Lieut. Col. John Frost and his survivors at Arnhem Bridge, where lack of courage or lowered morale could not possibly be considered. However, wargamers are notoriously lax on the question of prisoners, preferring to fight to the last man simply because it is difficult to provide on the wargames table the human inducement to surrender. It has to be made worthwhile in the broad sense for a wargame 'general' to surrender a portion of his force, although in real life the taking of prisoners is logical and highly realistic.

A possible inducement might be for each prisoner to require a 'keeper', in other words if twelve men surrender then it takes the same number of men out of the enemy's ranks to guard them. The tough airborne soldier, probably thinner on the ground than his opponents, could take out TWO men as guards for each prisoner. The guards take the prisoners back to a specified HQ or to the baseline, 50% of them being able to return to the battle – but the surrender has taken away valuable men at perhaps a crucial moment in the battle. Airborne troops might only need to provide one guard for every two prisoners; or it may be considered tactically expedient to refuse to accept (through dice-throws) surrender and shoot down the enemy – although this would seem to neutralise inducements.

The 1-for-1 (or 1-for-2) prisoner and guard method could be used in conjunction with an enforced surrender system, with double-strength attacking troops calling upon opponents to give in, who can fight it off by scoring 50% or more on the percentage-dice.

Street Fighting

A confused style of fighting extremely difficult to reproduce on the wargames table, but one that will occur when reconstructing airborne operations as the very nature of their operations often involved para- and glider-troops in street and house-to-house fighting, as they cleared the enemy from towns and villages and turned shattered houses into strongpoints, as at Arnhem. At the bridge and in the perimeter at Oosterbeeke, the paratroopers found that the advantages were always with the defender because tanks, artillery or aircraft cannot help the attacker once he has mingled with the enemy within a street or town, where a machinegun can command a straight stretch of street or crossing. Rifles and bayonets are an encumbrance and the best street fighters favour Tommy-guns, revolvers and Mills bombs, plus pick-axes or crowbars for 'mouseholing' from house-to-house. This was by far the best way of moving along an open street because not only did it provide a safe escape route but it also allowed unobserved movement of men and weapons from house-to-house. Once the first house was taken, holes were smashed in party walls, with the pick or crowbar wielder keeping well to one side of the hole in case the enemy fired through it, and then throwing a grenade through the hole into the next house before entering.

Street fighting was sometimes done with a force divided into two parties, one on each side of the street – the one on the lefthand side putting the enemy out of action in the first house on the righthand side, then No. 2 party made a quick dash to that house under cover of No. 1's cross-street fire. Then they gave covering fire across the street as No. 1 party attacked it. Both parties now directed their fire, always diagonally, at the next two houses on the right and left respectively, and so on right up the street. Another way was to keep close to the houses on the righthand side of the street so that their occupiers line of aim and vision was poor; the enemy in the houses on the other side of the street had to hold their rifles awkwardly to sight the advancing troops, unless they could shoot left-handed. Many paratroopers learnt to shoot like this, in the same way as switching the cue over to your left-hand at snooker.

When wargaming, if defenders are in undisputed possession of a house for two game-moves then they have had time to put it into a state of defence. This makes it much more difficult for the attacker to enter the house or fire at its occupants, who have barricaded and loop-holed doors and windows. Men firing from within houses in this manner should only take half casualties; otherwise defenders can only be fired upon when they expose themselves at windows as they take aim. A grenade or shell penetrating a defended room will probably clear it, but normal casualty deciding methods will apply. Apart from mouseholing, the only way to enter an enemy-held house is to break down the door, using a battering-ram, an axe or small explosive charges. Men breaking down doors must remain in the doorway for a complete game-move, they can be made to require 4, 5 or 6 on the dice to break the door down. Once this has occurred fighting can take place in the doorway, with some advantages given to the defender. If the invader is successful he enters the house and engages another defender within the house, this time on level terms. The simplest simulation is for both wargamers to throw dice, defender adding a bonus to his dice, the highest scorer being the winner; the loser is removed. If it is the defender the successful attacker passes into the house to take on another man; if the attacker loses, the defender gets on with his firing whilst awaiting another onslaught.

Fox-holes and Weapon-pits

It takes a man two game-moves to dig a fox-hole or weapon-pit large enough for his own tactical use, and it provides enough protection to at least halve his chances of becoming a casualty. This form of protection can be represented by moulding around the man (or men) a 5 mm (¼ in) high rough oval of earth-coloured Plasticene.

Mines and Minefields

Although there are few records of airborne troops laying mines to protect their defensive positions, it does not seem outside the bounds of possibility for both anti-personnel and anti-tank mines to be dropped in containers and laid by airborne troops, either in defence or as a means of disrupting enemy movement. They can be laid by paratroopers at a rate of an area 3 cms × 3 cms (1 in × 1 in) per game-move (when they cannot fight or take any other action) or by engineers at a rate of 4 cms × 4 cms (1½ × 1½ ins) per game-move. (These 'areas' link with the suggested method of simulating minefields).

In a defensive context, minefields may be in areas of airborne operations in-situ long before the particular action, so that their siting is known to the defenders who laid them. Their purpose was to slow down enemy attacks and allow time for defenders to concentrate or to 'channel' the enemy onto a 'killing-ground' registered by artillery and covered by fixed-line machine-guns. Often laid obviously, the psychological fear they inspired could cause the enemy to move his forces into such fire-covered positions. So the minefield was not the imagined unmarked and seemingly innocuous area of ground thickly sown with deadly explosive charges waiting to destroy the unwary person or vehicle who unwittingly ventured into its confines. Of course this happened when minefields lost their markers, or the darkness and stress of battle might cause panicky units to blunder into a field – but that could occur also to the unit that laid the mines! Generally, minefields were marked by low wire fences or perimeter notice boards bearing warnings in English or German, as the case might be.

A Sherman tank has its track blown-off when caught on a transparent plastic mine-field, its borders marked by pins to reveal its size and extent.

Although minefield concealment has been explained, no doubt the wargamer will wish to make attempts to hide these deadly weapons, and it can be simulated in a realistic manner with thin coloured transparent plastic sheeting used for covering books. Commanders are given a specified area of this material, say 20 cms × 20 cms (8 ins × 8 ins), from which to cut as many irregular minefield patterns as desired. Stuck onto the terrain, these green or sandy coloured transparencies are difficult to identify without close scrutiny and are sufficiently unobtrusive to allow occasional blunderings onto them.

There are two types of mines – anti-tank and anti-personnel. The effects of the former upon vehicles can be simulated by ruling that the tank has a 25% chance of getting through and a soft-vehicle a 10% chance – decided by percentage dice thrown each move that the vehicles in in the minefield. The effects of anti-personnel mines are represented by ruling that a man has a 20% chance of escaping, percentage-dice being thrown each game-move if the man is actually moving – as long as he stays still he is safe.

Smoke

A smoke-canister fired by a tank, gun or mortar will cover an area 15 cms (6 ins) square around the point where it landed; it lasts for three game-moves when there is no wind, two moves in a light wind and one move in a strong wind, drifting half its own area per move in the direction that the wind is blowing. Smoke from burning buildings or vehicles lasts for ten game-moves and extends in the direction that the wind is blowing at a rate of 5 cms (2 ins) per game-move.

Smoke is represented on the wargames table by grey-cotton wool about 7 cms (2½ ins) high, or by a series of different-sized realistically coloured cardboard shapes.

Conclusion

The suggested rules and conditions governing the table-top re-fighting of airborne operations are adequate to provide fast-moving and realistic wargames. There does not appear to be any existing commercial sets of rules solely relating to airborne warfare, so that the wargamer who is not fully satisfied with the suggestions in this book must invent and adapt until conditions conform to his personal interpretations.

It would be quite wrong to leave the subject of rules and the re-creation of warfare in miniature without mentioning Individual Skirmish Wargaming – a method of simulating combat on the table-top based on the individual soldier forming the basic 'unit' bearing a personal firing and fighting capability that can be affected by wounds and morale. Within each of the operations described in this book are a host of minor group or solo actions just crying out for table-top reproduction, as the marked degree of individual action and self-reliance in airborne warfare, with the combatants rigorously trained to high standards of physical fitness and weapon handling, causes it to be uniquely suitable for Individual Skirmish Wargaming.

Being a time-and-motion-study approach that re-creates in miniature Man's natural fighting faculties, the mechanics of play are simple and divide into phases representing five seconds of action, when each soldiers's intended movements and actions are recorded; then performed and the results calculated, with wounds and reactions taking effect.

Each figure represents a separate man and the rules encompass all the possible ranges of movement and actions of a reasonably fit human-being. Movement is measured in *distance* and action in *time*, with the latter covered by a five seconds period so that an element of delay is imposed to prevent the contestants becoming supermen. Basic rules and conditions are described in the book *Skirmish Wargaming* by Donald Featherstone (Patrick Stephens 1975) and, besides skirmishes in varying periods of military history, specifically considers a minor action of German paratroopers at Cassino.

WARGAMING CRETE AND ARNHEM

Every battle deşcribed in this book is worthy of simulation, each conflict lends itself remarkably to being re-fought in miniature on the wargames table, and the two operations selected for discussion in detail – Crete and Arnhem (Operation Market-Garden) – besides being among the most extensive, are also the best known and most evocative of all airborne operations. They are particularly suitable for simulation because, in their own way, each was a milestone in this form of warfare: Crete represented the zenith of German Fallschirmjager activities although spelling their doom, while British/Polish activities at Arnhem coupled with those of the Americans at Eindhoven and Niejmegen were the outstanding example of Allied airborne co-operation. Made-up of numerous smaller actions, each of these battles includes almost every aspect of the fighting and tactics employed by paratroopers and glider-troops, so that their reconstruction fully justifies the expense and effort of assembling miniature airborne armies, plus the necessary pre-planning and terrain construction.

The most obvious manner of reconstructing a battle is for the troops to perform precisely the same actions that they did on the day, take the same percentage losses, with the same success or failure. But this is an historical exercise and not a wargame, serving only as a demonstration of what occurred in the past, with hindsight ignored. It is preferable to follow the original course of events reasonably closely allowing some leeway but not overstretching the imagination. Although legitimately seeking to reverse the historical result, too many liberties should not be taken or the battle will become merely a wargame played for its own sake. Every aspect of the conflict should be considered in its correct context and in chronological order, and the fluctuations and fortunes of war considered without radically departing from the military possibilities of the day. It is vital for the wargamer to consider what MIGHT have happened, in conjunction with what DID occur.

Scaled down to a suitable area and frontage, the size and topographical features of the real-life battlefield should be reproduced on the wargame table, although it is pointless to construct a terrain closely resembling the area around Maleme airfield in Crete, for example, and then allow paratroopers, New Zealanders, tanks and artillery to generally mill around in a manner that bears no relation to the original conflict.

Realistic scaling down of numbers is important because if the battle is to bear anything more than a titular resemblance to the original the table-top armies, both in numbers and types, must represent an accurate proportion of the real-life forces. When re-creating these two epic battles it is not essential to think of the actual numbers of men, better to consider the units and formations that are clearly defined in the accounts. Thus, instead of one man on the wargames table representing 20 men in real-life, allow a table-top unit (even if it is only 20 men) for each formation historically present. This is facilitated by the manner in which the battles were broken up into a number of smaller actions, each requiring to be fought separately on the wargames table, allowing ALL available model soldiers to be used for each game. This is vastly superior to saying . . . 'I have got 500 British and American paratroopers to be divided into the various units of the British 1st Airborne Division at Arnhem, the 101st U.S. Airborne Division at Eindhoven and the 82nd U.S. Airborne Division at Niejmegen'. Such a method broken down into battalions means that each will consist of about two figures! If *all* 500 are used to represent single phases of the battle in turn, then formations of far more realistic sizes can be fielded.

Ideally, varying accounts of a battle and the events leading up to it should be read to achieve a familiarity with the factors that influence its trend and pattern. However, the accounts given here are sufficient to put the wargamer in the picture, to reveal the objectives of the battles, and to arouse an awareness of each commander's intentions and an understanding of his tactical plans. After reading about the battle, analyse it and seek out those aspects that led to victory or defeat; consider each phase carefully and seek possible alternatives (in the shape of Military Possibilities) to the historical trend of events. For example, what would have occurred during the German invasion of Crete had Hill 107 at Maleme not been abandoned?

Over and above the normal provisions of the conventional wargames rules given in this book, other factors must be considered if an accurate and realistic simulation is to be achieved. Perhaps the most potent and credible of variants are Military Possibilities – controlled and logical alternative courses of action that, had they been taken at the appropriate course of the battle, might well have caused a complete reversal of its result. Not excuses for indulging in whims and fancies nor for diverting events merely 'to see what happens', Military Possibilities are agreed courses of action that sometimes result in a more reasonable and credible result than actually occurred!

The throw of a dice is the simplest means of simulating War's fluctuations of Fortune; allied to the use of Chance Cards, they can form the LUCK aspects of Military Possibilities.

Controlled Military Possibilities can either radically alter the historical course of a battle or affect only its minor aspects, although even an apparently simple tactical 'twist' can influence the eventual outcome. Historical battles abound with them – on the wargames table they bring interest and colour, depending upon the ingenuity of the wargamer.

Chance Cards

There were many instances and situations at both Crete and Arnhem to provide suitable occasions for the use of Chance Cards. Their rulings intoduce pleasant or unpleasant factors materially affecting the battle – even its result – by posing tactical, physiological or psychological eventualties that have to be tackled by the commander drawing such a Card. For example, had the glider bearing Gough's jeeps intended to dash for the Arnhem bridge NOT been destroyed, then the whole outcome of the battle might have taken a different course – a Chance Card could settle this one way or the other. However, Chance Cards should be used sparingly and with discretion, or the historical aspects of the battle could be unnaturally distorted. They consist of a set of cards peculiar to each situation, marked with varying conditions and alternatives that must be followed by the drawer of the card.

Time Charts

Both battles under review were radically affected by the clock, the anticipated time of arrival of reinforcements and supplies being of vital tactical importance. These battle factors are programmed beyond dispute by using Time Charts, which co-ordinate manoeuvres and stages with game-moves as they occur.

Crete and Arnhem, fought as wargames, are 'sectional' battles; with German East, West and Centre Groups fighting quite separately from each other, while Operation Market Garden saw the 82nd and 101st U.S. Airborne Division's activities south of Arnhem, taking place simultaneously with Frost's defence of the bridge and attempts from the northern landing and dropping-zones to join him. A Time Chart is essential to keep check of such simultaneous and programmed situations taking place on a map, with different forces moving along various routes or attempting movements to bring troops onto the table-top battlefield at intermediate stages of the conflict.

Communication with detached parts of a force require commanders, perhaps unaware of their exact location, to send messengers, whose progress must be recorded on the Time Chart so that their exact time of arrival is noted. The unit to whom they are bringing orders does not react until those orders are received – non-arrival or delay provides Military Possibilities that can realistically alter the course of the historical conflict.

Pre-Planning

It has to be decided whether wargaming these battles is to be done as a series of single battles eventually co-ordinated to produce an overall result, or as a 'campaign' ideally involving a number of wargamers. They will represent the various commanders on the spot, either under the direction of an overall commander or, through lack of communications, forced to make their own decisions during the course of the battle, as historically occurred.

If fought as a campaign, the wargamers will require at least one fairly extensive 'briefing' meeting to allocate roles and consider the practical aspects of setting up a relatively extensive wargames operation. As both battles consisted of a number of simultaneous small actions, it is preferable when wargaming them to reproduce such conditions by each battle being fought on a separate wargames table. Thus, Wargamer 'A' will set up the area around Maleme airfield; Wargamer 'B' the Retimo area; while Wargamer 'C' is responsible for the Heraklion area. The wargames, unlikely to be concluded in a single evening, will progress until conclusions are reached, with participants pledged not to reveal the progress of their game to the wargamers handling other areas.

A Co-ordinator will mark up a large map of the entire area of operations (copied in larger scale from maps in this book, or a commercially produced map). Each wargamer will require a large-scale map of his immediate area of operations, progressively marked with troop dispositions and movements – these sector maps are drawn 'wargames-style', with topographical features condensed to fit table space and unnecessary features omitted. Holding regular meetings to discuss progress (without revealing successes or failures until the appropriate time) produces a sustained atmosphere of interest and excitement.

If this system of re-fighting the campaigns is not possible, it is necessary to decide the order of fighting the actions, when chronological factors do not apply. A knowledge of the result of preceding battles will not greatly detract from the interest or intensity if the entire campaign is considered on a 'league table' basis, with points allocated to each battle and totalled at the conclusion of the campaign to decide the overall winner. Otherwise, it is not beyond the scope of intelligent fair-minded wargamers to generally assess results to reach a conclusion.

All of these smaller actions are suitable and worthy of reconstruction on the wargames table, and wargamers can choose which to reconstruct without necessarily embarking upon a prolonged campaign, although this will eliminate the most compelling element.

With hindsight, the German invasion of Crete falls neatly into a pattern with the action at Maleme all important and German activities at Heraklion and Retimo merely containing local garrisons – when Maleme was captured, the results of the other two areas were unimportant. On the other hand, Operation 'Market Garden' required an essential linking-up between airborne forces seeking to capture a series of bridges while a relieving force battled to contact them. Thus, it is more essential to re-fight Market Garden' as a campaign with an overall result.

An early decision has to be made whether the reconstruction is to be a reasonably accurate simulation or a wargames re-fight. With so much information available on formations, dispositions, tactics, objectives, timings, etc., etc., it would seem easier to accept the known facts and faithfully reconstruct these historical actions, allowing selected Military Possibilities to provide variations and uncertainty at appropriate tactical points.

On the other hand, if these actions are being re-fought to ascertain whether alternative tactical planning will produce more successful results, more pre-planning organisation will be required, repaid by many intriguing possibilities. At his disposal, the wargamer will have scaled-down numbers of aircraft to carry in any fancied order of men, weapons and equipment in varied lifts, perhaps in other places. If the reconstruction is to bear a noticeable similarity with real-life events and be more than a wargame the table-top terrain must closely resemble that of real life, with the same dropping and landing-zones, and with scaled-down numbers that approximate to those in the original battle. Changes should consist solely of variations in attacker's aircraft capacities and times of arrival; and the defender has similar leeway in his dispositions.

It being impossible to pin-point the exact point at which each individual soldier or glider landed drops and landings will be made in accordance with the methods described in the appropriate section of this book, being scattered on and about the zones. Those are the general aspects of wargaming Crete and Arnhem – here are more specific features of those two campaigns.

THE BATTLE FOR CRETE, 20/30th MAY 1941

It has been asserted that the invasion of Crete could have been beaten at the very outset had defenders, rather than limiting themselves to purely defensive measures, attacked the paratroopers immediately and energetically as they landed. If this is done on the wargames table it will obviously result in a different type of battle. Therein lies an intriguing historical variation that could lead to the discovery of what would have occurred had alternative tactical policies been followed. Otherwise, begin each battle with the defenders in their real-life positions and, when it is known to have occurred, subject them to an initial air bombardment before paratroops and gliders come in; then assess casualties and test morale perhaps repeating this when the air drops begin.

Despite low fire-power and other limitations, all defenders seem to have fought well at Crete. But it is not unreasonable to assume that while all defenders are equal, some are more equal than others. For example, the Black Watch, the New Zealanders and Australians were obviously higher quality troops than the poorly armed and under-strength Greek battalions and Cretan Police, gallantly as the latter appeared to have fought. Some form of bonus for fighting and morale should be given to the better units, or a diminished valuation bestowed on the unfortunate locals.

Everything points to the fact that German success on Crete was due to high standards of leadership, courage and initiative at all levels, particularly among the paratroopers themselves. This must be reflected in the rules, and where there are stipulations concerning the effects upon troops when bereft of officers, they must be considerably lessened for the Germans. If the numerically inferior Fallschirmjager are to have any chance of repeating their historical success, they MUST be classed as ELITE troops, with bonuses for both fighting and morale.

Simulated by suggested methods the initial German landings of paratroopers and gliders might possibly occur in slightly different places to those in real-life i.e. Major Scherber's 3rd Bn. falling among the New Zealand positions and losing 400 out of 600 men within 45 minutes, may not be so unfortunate on the wargames table, perhaps even dropping on or close to the beach, their intended area.

On numerous occasions German paratroops either failed to find their weapon-containers, or were prevented from reaching them by the fire of the defenders conscious of the importance of the cylinders.This fact can either be made known to defenders before the battle commences, or revealed to them through Chance Cards.

It is also by means of Chance Cards that the confusion arising from the shortage of radios, leading to an almost complete lack of communication between Freyburg's HQ and his formations, is simulated.

The small German seaborne invasion can be a paper transaction with the odds weighed against the Germans to represent their eventual destruction by the Royal Navy. However, Chance Cards or Military Possibilities can be used to decide whether a certain percentage of ships survive and land their cargoes of light armour, artillery, supplies, heavy equipment or men. On 28th May, the Germans used against Campbell's Australians at Retimo two tanks that had been landed by sea. It is arguable that if the Germans could get two tanks ashore they might have got more – this can be decided by Chance Cards and an interesting beach-battle could be set up with light German tanks coming into action straight from their landing-craft.

A typical Military Possibility occurs at Retimo where the 2nd Regiment's attack on the airfield might have succeeded if the 3rd Bn. had aided them by leaving Retimo town and attacking the heights overlooking the airfield. This can be simulated by giving the 3rd Bn. the chance to take this action, with the odds weighed so that they are likely to follow their real-life course and fail to accept the opportunity.Another Possibility could allow Heilmann's 3rd Bn. to come in WITHOUT using the plane-load type delivery that was so unsuccessful as to cause them to be almost wiped out.

Coming in at Heraklion airfield, Lift two encountered six tanks and 16 Bren carriers, a surprise that must be reproduced on the wargames table, with the wargamer handling the Germans having no idea of what is in store for him. Nevertheless, here the 1,000 German paratroopers in two groups displayed such aggression against three times as many defenders supported by tanks and light weapons, that 14th Australian Infantry Brigade were thrown on the defensive. It was here that paratroopers built a landing strip behind a hill and flew in supporting airborne artillery. These two aspects both held great promise for a wargame, with the Australians forced to adopt a defensive policy and the aggression of the Germans encouraged by 'local' rules. Military Possibilities or Chance Cards can control the arrival of the supporting artillery.

Lift Two was late leaving Greek Airfields, so that its preliminary air bombardment was over well before it came in, allowing the defenders to recover and inflict heavy casualties upon the new arrivals. The degree of time elapsing between the German preliminary air bombardment and the parachutist's arrival can be controlled by a Military Possibility or Chance Cards. By these means the bombardment, occurring in a tactically correct manner immediately before the landing, will give the paratroopers an immense advantage. General Student was fortunate in amassing a parachute reserve with part of his anti-tank battalion which had been held back, and with troops delayed by mechanical failures in both air lifts. A Military Possibility could rule that this reserve did not exist, because the factors which caused them to be available did not occur. Bereft of these reinforcements, the Germans in Crete might well have failed.

At one stage the Germans had almost surrendered the initiative and, had they been defeated at both Retimo and Maleme, the invasion would have failed. At Heraklion, Brauer's 1,000 men contained a strong Allied force which could have usefully reinforced the Australians at Retimo, or the New Zealanders at Maleme. In the event, at Retimo Campbell's two Australian battalions were completely triumphant, but the campaign was virtually lost at Maleme through the abandonment of vital Hill 107. These events can be linked and by using such Military Possibilities as the New Zealanders NOT abandoning Hill 107, history realistically reversed at Crete. At this time the wargamer controlling the German forces (representing General Student) has to take the decision whether to send in regardless of losses his remaining supply-aircraft, two parachute battalions, and the Mountain battalion to establish a firm foothold at Maleme.

Freyburg's abortive counter-attack of the 22nd can be simulated and represents an interesting 'single-evening' wargame.

As a map/wargames table campaign, the Allies fighting withdrawal across the White Mountains to the Port of Spakia on the south coast contains the elements of some interesting rearguard actions and Individual Skirmish wargames.

In real-life, Crete was touch-and-go and, with hindsight, undoubtedly could have been a disastrous set-back for the German airborne forces – in the event, their immense losses marked the virtual end of their true role. An operation so finely balanced is tailor-made for table-top reproduction, if only because a reversal of the historic result is not outside the bounds of possibility.

OPERATION 'MARKET GARDEN' (ARNHEM)

17/24th SEPTEMBER 1944

Reproduced on the wargames table, the 'MARKET' part of the Operation – the actions of General Brereton's 1st Allied Airborne Army – could be a mammoth affair suitable as a club-activity spread over a winter season. Known details can be adapted to provide an accurate historical re-creation varied by judicious Military Possibilities, or quite different tactics can be used in investigating the possibilities of achieving success on the same fields, but by alternative methods.

The Operation conveniently divides into the three principal actions of the two U.S. and the single British Airborne divisions, being again sub-divided into a number of smaller operations, mostly highly suitable as wargames. On larger scale, the wargamers representing the commanders of U.S. 82nd and 101st

Airborne Divisions and British 1st Airborne Division can be given the historical orders and objectives and left to make their own plans.

At even a reasonable scale of 1 figure = 20 real-life men, the 25,000 strong 1st Allied Airborne Army requires 1,250 figures. This is not extraordinary using Airfix or 1:300 scale figures, with participating wargamers possibly each contributing the formation they are to command. By splitting the operation into its separate actions and using all available troops for each, more respectable numbers are available and no figures lie idle awaiting their part in the wargame. The same provisions apply to the German defenders, detailed in the account of the battle.

Hindsight reveals the unfortunate existence of unexpectedly strong German forces in the area – it will have to be decided whether to position them in their historical sites (depriving them of the valuable element of surprise) or to perpetrate a double-surprise upon the Allies by positioning the Germans elsewhere, when the Allies are anticipating that they will be in their historical areas! German reactions to the landings will undoubtedly be accompanied by a degree of confusion – simulated by Chance Cards covering speed of reaction; misadventures of messengers; quality of commanders and effect upon subsequent actions, etc., etc.

It is now well known that in the early stages a complete set of Allied operation orders were captured by the Germans. There are so many readily available details of objectives, formations, etc., etc., that this is hardly relevant *unless* the wargame is being fought on the understanding that the Allied commanders are using *different* tactical plans to those of real life. Then it might be advantageous to allow a copy of the wargames operation orders to fall into enemy hands, without the wargamers representing the Allies knowing about it!

Within the framework of the operation are contained examples of many different types of wargame. A most unusual simulation is the epic battle of Frost's small force against overwhelming numbers of German infantry supported by heavy tanks, S.P. guns, artillery and aircraft. The table will need to be fully patterned with houses, gardens, roads and the Bridge itself, its northern approach at one end of the table, the scene of repeated German attempts to rush across in the face of the defenders anti-tank guns. On the three other sides of the table there are attacks from house-to-house with flame throwers, tanks, SP guns, etc. When considering wargaming the battle of Arnhem, this is one picture that primarily comes to mind!

Another is the fighting around the perimeter, with bayonet-point actions, snipers in the trees, armoured cars and tanks prowling around in the woods and hotels changing hands time-and-time again after fierce melees. This should prove irresistible to the wargamer, but he will find it complicated fighting that requires strict rules and a great sense of give-and take! Target-times (i.e. allotted number of game-moves) are required by both defenders and attackers to reveal success or failure. It is advisable that these wargames be controlled by strong umpires because much will depend upon the achievements of concealed PIAT's on German armour, and few rules can satisfactorily allow for concealment. Much will depend upon the goodwill of the wargamers and the strength of decision of the umpire!

The delaying action fought by Krafft's SS Training and Depot Bn. allowed the 9th SS Panzer Division to get into position. This action, carefully programmed on Time Charts, possesses all the background and purpose for an excellent wargame.

Then there are the attempts of Hackett's 4th Parachute Brigade and the remnants of the Air Landing Brigade to reach Frost at the Bridge. Both actions can be linked so that the success or failure of the relief attempt depends upon how long Frost can hold out. Otherwise allocate a specified number of game-moves on the Time Chart to provide both the relief force and the Germans with a target at which to aim.

Containing almost every stimulus and ingredient for good wargaming, Operation 'MARKET' reveals stirring scenarios for Individual Skirmish games such as the attempt by Lieut. Grayburn's platoon to rush Arnhem Bridge from the north, and subsequent feats that won him a posthumous Victoria Cross. The same award was won by L/Sgt. Baskeyfield within the perimeter at Arnhem where he destroyed two Tiger tanks and an SP gun at 100 yards range with a 6 pdr anti-tank gun. And a fine tank-hunting Individual Skirmish is Major Cain's winning of the Victoria Cross by immobilising a Tiger tank with a PIAT.

From the moment of the landings, Military Possibilities abound, beginning with the 35 gliders that failed to make the landing-zone on the Arnhem operation so that the armoured jeeps of the Reconnaissance Squadrons did not arrive. Had these vehicles been available to make their scheduled dash for the bridge it is quite likely that Frost's epic defence would not have been necessary and events might have taken a different course. But, in the long run it might not have made much difference unless XXX Corps could break through to reach them.

Perhaps the greatest Military Possibility arising from Operation 'Market Garden' lies in the fact that on the first evening the British 1st Airborne Division and the German 9th SS Panzer Division were both moving towards the Arnhem Bridge, the Germans aware of their enemy's general position and intention, whereas the British Airborne forces did not even know that the German Armoured formation existed!

Long before this, a Military Possibility concerns the main tactical flaw in the British plan i.e. the location of the dropping and landing-zones of the 1st British Airborne Division. The real-life objections to landing-zones close to the Arnhem Bridge may not impress the wargamer who might feel inclined to take a tactical chance and establish his dropping-zones where originally intended.This alters the complexion of the battle, but provides such an interesting exercise in what might or could have occurred that it is probably well worth considering.

The allocation of first-day lift capacities – the Americans devoted more than three-quarters to foot soldiers whereas the British allocated only half to infantry and took in all their vehicles and heavy gear – can be the subject of Military Possibilities. The wargamer commanding these formations might, with hindsight, wish to change these capacities, which could alter the principal features of the operation but is an interesting exercise. General Urquhart was separated from his force for a vital period – a Military Possibility could prevent this occurring and 'local' conditions could be applied to reflect the value of his presence.

The British 1st Airborne Division were plagued throughout the battle by lack of or faulty communications, which can be reproduced by Military Possibilities or Chance Cards.

The adverse weather conditions that delayed the Polish Brigade landings can be reflected by Military Possibilities, but may well be more affected by faulty communications.

Directed by observations posts within the perimeter, British guns fired from South of the river – an excellent opportunity to reproduce and practise off-the table artillery firing.

The British paratroopers marching from their distant dropping zones to Arnhem itself will take place on a map, until such time as opposition is encountered when operations are transferred to the wargames table. Similarly, the activities of 101st U.S. Airborne Division will require map-moves, with occasional forays on the table when contacts are made. This American part of the reconstruction will require the provision of elements of the Guards Armour Division as contact was made between the Americans and the relieving force.

Gavin's 82nd U.S.Airborne Division had a series of adventures which provide material for at least half-a-dozen good wargames, beginning with some Individual Skirmishes as paratroopers dropped right onto German anti-aircraft batteries around the dropping-zone. Similar opportunities occur with the platoon of 504th Parachute Regiment attacking bridges over the Maas-Waal Canal and knocking-out a flak tower with a bazooka, turning its 40 mm gun against its former owners.The capture of the Grave Bridge and the cutting of the wires of explosive charges forms a wargame with a notable objective.

The 508th U.S. Parachute Regiment, battling desperately for the Honinghutie Bridge, were aided by a patrol of 504th Regiment creeping across the bridge and attacking the German defenders in the rear. Using map-moves or some form of concealment, such an occurrence is an interesting side-issue to the overall wargame. Then there was the fierce fighting by 508th Regiment to eject Germans overrunning landing-zones where the 82nd's gliders were due to arrive. Fortuitously they were delayed two hours but still landed under fire – Chance Cards can decree that the gliders land either earlier and take greater losses, or later and escape entirely.

On 19th September, German troops were ferried across the Rhine to reinforce formations fighting the Americans south of the river. This could provide a wargame or an interesting aspect of the campaign with Chance Cards or Military Possibilities enlarging the operation.

A fine wargame is the American assault boats attack across the Waal, simultaneous with attacks on both ends of the Nijmegen Bridge. Smoke was used to cover the crossing and can be reproduced on the table-top, where the wargamer will need to construct a simulation of the bridge itself in the centre of the table with defenders AND attackers on BOTH sides. Also required are elements of Horrocks relieving force, as Guard's tanks (including Sherman Fireflys with 17 pdrs) formed an essential part of the attacking force. This armoured action, most exciting to read about, could be memorable on the wargames table!

The complicated timings and programming of events – both Allied and German – during this extensive operation makes apparent the need for very comprehensive Time Charts if the necessary degree of control is to be maintained.

The reasons for the failure of Operation 'Market Garden' provide the background to a table-top wargames campaign, or fruitful material for attempting to reverse historical events by using different tactics. The overall battle included many notable ground actions that will go down in military history – that of the 82nd U.S. Airborne Division was almost classical – and reveal almost every aspect of World War Two combat styles and tactics. This glorious but tragic military operation is truly suitable and worthy for table-top reproduction.

APPENDIX ONE

THE COMPOSITION OF AIRBORNE FORMATIONS

Airborne Divisions varied between nations –the Germans, British and Americans patterned theirs on infantry divisions between 12,000 and 18,000 strong, composed of three brigades (or regiments) plus such support and service units as artillery, engineer, medical, ordnance, provost, reconnaissance, signals, supply, transport and workshop. These divisional troops were scaled-down in airborne divisions.

In German airborne divisions all three regiments were parachute. Glider-tasks were undertaken by parachute troops, although at one stage a parachute-trained Assault Regiment was raised specially for the glider assault role.

British and American airborne divisions were formed of two parachute brigades (regiments) and one glider-borne. Towards the end of World War Two, U.S. Airborne Divisions in Europe were raised by a third parachute regiment. Each brigade or regiment was formed of three battalions with such attached divisional troops as anti-tank, signals, light-aid detachment and light anti-aircraft.

Glider-borne infantry and parachute battalions consisted of three or four rifle companies, a support or heavy weapons company and a headquarters company. Rifle companies varied in strength – in the British Army (where support weapons were not grouped at this level) they consisted of 120 men; in the American Army (where the company was a fully self-contained fighting force) a company was 200 men. Light support weapons were provided within companies; medium mortars, machine guns and anti-tank guns were usually grouped in support companies. Perforce, airborne battalions had to scale their support to what could be delivered by air or carried by the soldiers themselves, or on the few available vehicles, and to what could be kept supplied with ammunition from the air.

Companies were formed of three or four platoons, each divided into three or four sections or squads who formed a 'dropping unit' of a sergeant, a corporal and eight men (U.S. 10 men), armed with rifles, bayonets, grenades, sub-machineguns and a light machinegun. When they became available, short-range anti-tank weapons were issued to sections and platoons – the Americans had the 'bazooka' rocket-launcher, the British the Projector Infantry Anti-Tank (PIAT), and the Germans the Panzerfaust expendable rocket-launcher. It was customary for parachute sections to substitute ordinary rifles with sub-machineguns, and telescopically sighted sniper rifles were used.

The German and Allied Airborne Divisions were broken down in the following manner:

GERMAN AIRBORNE DIVISION

Airborne Division

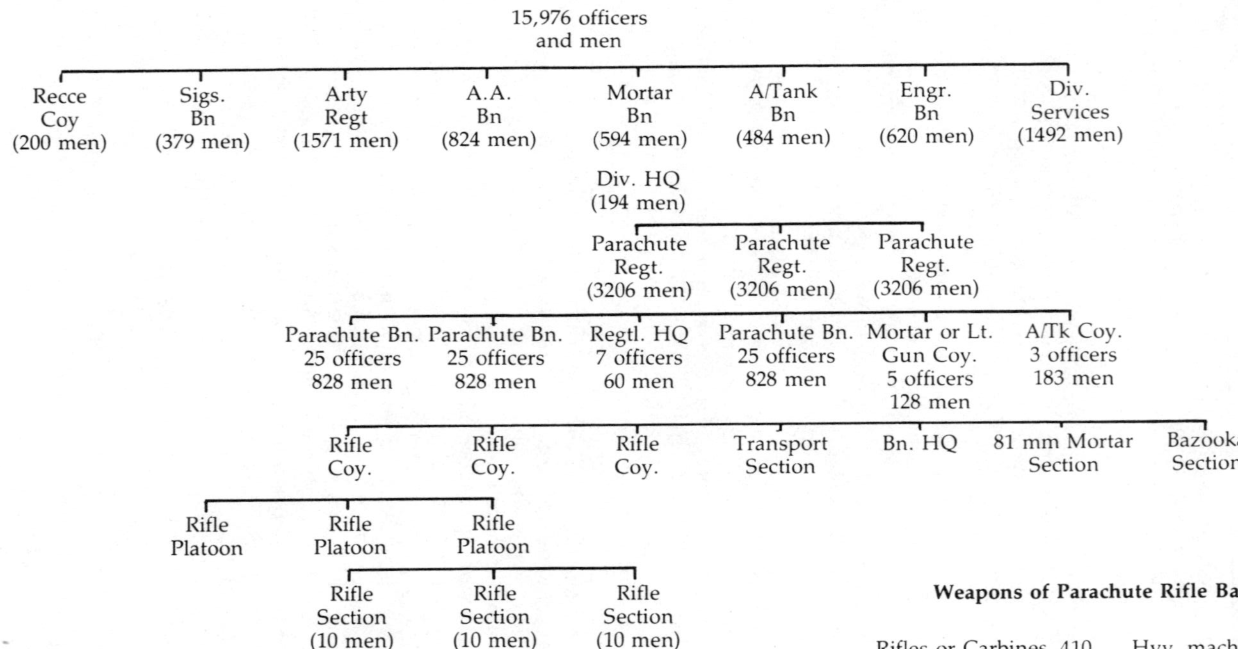

Weapons of Parachute Rifle Battalion

Rifles or Carbines	410	Hvy. machineguns	8
Pistols	257	81 mm Mortars	13
Sub-machineguns	214	Bazookas	54
Lt. machineguns	66	75mm Lt. Guns	2
		75mm A/Tk Guns	3

AMERICAN AIRBORNE DIVISION

Division
(12,799 men)

Div. HQ

Parachute Regt. (5,184 men) | Parachute Regt. (5,184 men) | Glider Regt. (5,184 men) | Engr. Bn. | Artillery Bn. (Incl. A/Tk, A.A., Field) | Signal Coy

Rifle Bn | Rifle Bn. | Rifle Bn. | Bn. HQ | Weapons Coy. | Arty. Coy. | A/Tk Coy.

Rifle Coy. | Rifle Coy. | Rifle Coy. | Weapons Coy. | A/Tk Platoon

Rifle Platoon | Rifle Platoon | Rifle Platoon | Weapons Platoon

Rifle Sqd. 12 men | Rifle Sqd. 12 men | Rifle Sqd. 12 men

Weapons Allocation

Div. Arty Bn. were armed with 105mm Howitzers Regtl. Arty Coy. had six 105mm Howitzers.

Regtl. Weapons Coy. — 6 81mm Mortars
8 .30 LMGs
3 .50 HMGs
7 Bazookas

Regtl. A/Tank Coy. — 12 57 or 75mm A/Tk guns
1 .50 HMG
4 .30 LMGs

Bn. Weapons Coy. — 3 81 mm Mortars
2 .50 HMGs
4 .30 LMGs
6 Bazookas

Bn. Weapons Platoon — 3 60mm Mortars
1 .50 HMG
2 .30 LMGs
3Bazookas

Riflemen armed with — Browning (B.A.R.) Automatic Rifles
.30 Garand semi-automatic rifles
Thompson Ml sub-machinegun
.45 M3 sub-machinegun
Pistols
Grenades
Bazookas

BRITISH AIRBORNE DIVISION
(Based on standard Infantry Division)

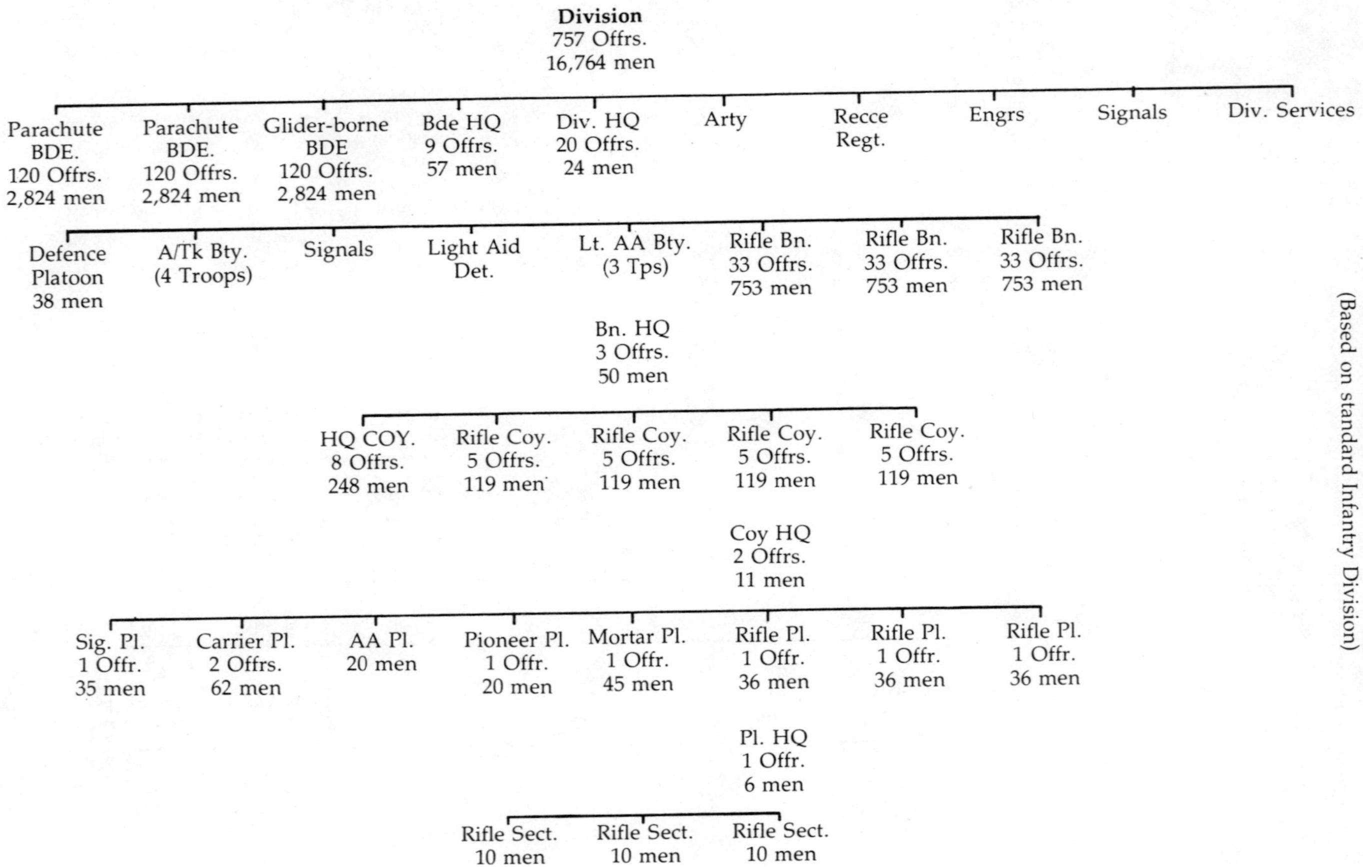

WEAPONS ALLOCATION (Battalion Level)

	Pistols	Rifles	SMGs	Brens	Piats	2″ Mortars	3″ Mortars
Bn. HQ	11	42	2	1			
Sigs. Pl.	4	32			1		
Carrier Pl.	2	58	4	13	1	4	
A.A. Pl.	1	10		4 (Twin)	4		
Pioneer Pl.	1	21					
Mortar Pl.	4	42			2		6
Rifle Pl.	1	33	3	3	1	1	
Pl. HQ	1	6			1	1	
Rifle Sect.		7	1	1			

This is the allocation for an infantry battalion; it was customary for parachute units to acquire sub-machineguns and pistols in place of rifles.

APPENDIX TWO

SETTING-UP REALISTIC BATTLEFIELDS

When reconstructing an actual battle the topographical features and the dimensions of the terrain must closely resemble the historic field, otherwise what occurs upon it bears only the most coincidental resemblance to the original battle. The best size for a wargames table is whatever suits you, but it should not be more than two metres wide unless you have very long arms! The author's well-worn table measures 8 feet by 5 feet.

Wargames terrains should only include those parts of the field over which fighting actually occurred, or those parts which have a real bearing on the realistic simulation of the battle. Otherwise, all the model soldiers will be crowded into perhaps half the table while the remainder is unused.

The easiest manner of producing an undulating surface is to stretch a green cloth or a suitably coloured plastic sheet over mounds of books, slabs of polystyrene or pieces of wood. All hills and slopes should be so graded that model soldiers can stand-up on them. Strips of suitably coloured cloth, paper or Fablon are stuck onto the surface to form rivers and roads, or they may be painted on the plastic sheet with poster paint.

The most realistic wargames terrain is constructed on a sand table, moulded into hills, valleys, sunken roads, river-beds, trenches, etc., etc. However, sand tables are very heavy and messy, and pleasing terrains take a long time to build and colour. They are discussed in some detail in the book *Wargames* and the booklet *Wargames Terrain* (published by the author).

Presenting a stylised appearance, slabs of wood or polystyrene, placed upon each other, form stepped-hills providing readily definable contours and an ideal surface on which to stand model soldiers.

Few stretches of land completely lack vegetation; the appearance of a battlefield is made more convincing by clumps of trees, bushes and scrub. Having to be accessible to model soldiers who may be fitted three or four to a stand, space must be left for manoeuvering. This is achieved by painting an irregularly shaped piece of hard-board a darker colour than the table-top, and glueing a few trees around its edge – the whole area represents the wood. Trees and hedges can be made from lichen-moss stuck on twigs, or purchased ready-made from model railway and hobby shops; stone walls, bridges and rail fencing are purchased, made from balsa wood or Plastikard; broken pieces of polystyrene, painted grey, make ideal crags and rocky outcrops.

Plastic kits of houses can be purchased and buildings may be made from card. To arrange troops within the houses they are defending is difficult and many wargamers remove from the table the number of troops claimed to be in a particular house, keeping check on them by laying a small V-shaped numbered ticket across the roof-ridge using different numbers as casualties reduce the number of men in the house.

The Bellona Battle Game Scenery range of ready-made scenery and landscape models (Micro-Mold Plastic, 1 Unifax, Woods Way, Goring-by-Sea, Sussex, England) offers a large variety of P.V.C. terrain-pieces which are inexpensive, realistic and effective.

There is a deplorable tendency among wargamers to lavish time and attention on model soldiers, while neglecting the table-top terrains over which they fight. Enjoyment of a wargame varies in relation to the degree of realism of the miniature battlefield, so that it should have a pleasing and authentic appearance. Any of the battles described in this book can closely resemble their historical counterparts but, even if history is ignored, a realistic terrain will ensure that the wargame bears *some* resemblance to the real battle!

glider in
d scale from a
kit by Italaerei
urtesy of Revel
Limited).

Model of a 1:72nd scale Douglas AC47 (Dakota) made from an Airfix kit (photograph by courtesy of Airfix Products Limited).

Model of a Junkers Ju52/3M Transport Aircraft made from a kit – the model depicted has been given floats but the picture on the box cover shows it in the use with which we are most concerned (by courtesy of Airfix Products Limited).

Model of a 1:72nd scale Short Stirling made from a plastic kit (by courtesy of Airfix Products Limited).

APPENDIX THREE

SOURCES OF SUPPLY

It is fitting that a style of warfare so well-suited to the wargames table should be backed by exceptional sources of inexpensive and readily available model soldiers and equipment. Almost every type of soldier, gun, tank, aircraft, glider and weapon mentioned in these pages can be obtained in scales ideal for our purposes. The most severe limitation to amassing large armies of model soldiers is the time taken to paint them –the 'airborne' wargamer is fortunate in that some figures are so small in scale as to make painting a formality, or else moulded in plastic the same basic colour as the man's uniform – only flesh, equipment and weapons need painting.

The 1:300 scale figures, vehicles and aircraft made by HEROICS AND ROS FIGURES (36 Kennington Road, London SE1) are very cheap yet each model is perfect and easily recognisable as what it represents. Their minute size permits large forces to be manoeuvred on averaged-sized wargames tables. The range includes paratroopers and infantry, airborne artillery, Tetrarch tanks, the principal gliders and aircraft of both Allies and Axis.

The ubiquitous AIRFIX range of OO/HO scale plastic model soldiers, vehicles and equipment provides paratroopers and infantry, a few pieces of artillery and some aircraft of America, Britain and Germany. A similarly inexpensive range of plastic figures in this scale are those made by the Italian firm of ATLANTIC, now available in Great Britain. Closely resembling Airfix figures in many of their stances and types, there are mortars and anti-tank guns, etc., that can be easily converted to fit in with existing Allied or Axis armies.

From HINCHCLIFFE MODEL FIGURES (83 Wessenden Head Road, Meltham, Yorkshire) comes a fine metal 1:72nd scale 17 pdr anti-tank gun and crew, with a small range of figures suitable for World War Two wargaming.

Peter Laing (Minden, Sutton St. Nicholas, Hereford, HR1 3BD) can provide a growing range of 15 mm metal figures of World War Two infantry.

H0/00 scale plastic World War Two British paratroopers (photograph by courtesy of Airfix Products Limited).

The wargamer who fancies Individual Skirmish fighting will find the inexpensive AIRFIX 1:32nd plastic figures to be most suitable for his purpose – the range includes paratroopers and infantry of all three countries plus German Mountain troops, and Australians suitable for a Crete re-construction. BRITAIN'S plastic/metal DEETAIL range of World War Two figures, and those by TIMPO are also suitable for this type of wargaming.

The Italian firm of ITALAEREI (British Agents – Revell) produce 1:72nd scale plastic kits of the gliders Waco CG-4a; Gotha GO242 and Horsa. If considered necessary, plastic kits of most of the carrier and tug aircraft can be obtained from various well-known makers in 1:72nd scale.

No doubt there are other manufacturers, mostly Continental or Japanese, who make figures, vehicles and aircraft that can be

H0/00 scale World War Two German paratroopers (photograph by courtesy of Airfix Products Limited).

H0/00 scale World War Two German Mountain Troops – who acted as airborne troops in many theatres – (photographs by courtesy of Airfix Products Limited).

used for airborne wargaming – it pays to keep an eye on advertisements in magazines such as WARGAMER'S NEWSLETTER (Tradition , 5a-5b Shepherd Street, Mayfair, London W1Y 7LD); MILITARY MODELLING, BATTLE AND AIRFIX MAGAZINE (newsstands). These publications often include articles and photographs associated with airborne warfare, as does WAR MONTHLY (newsstands) and WW2 JOURNAL (218 Beech Street, Bennington, Vermont, 05201, U.S.A.).

BIBLIOGRAPHY

Beaumont, Roger A. — Military Elites (1976)
Blackwood — Soldiers' Battle Tales from 'Blackwood' (1968)
Edwards, Roger — German Airborne Troops (1974)
Ellis, Chris — German Military Combat Dress 1939-1945 (1973)
Ellis, Chris and Chamberlain, Peter — Handbook on the British Army 1943 (1943)
Feist, Uwe — Fallschirmjager in Action (1973)
Foss, Christopher F. and Gander, T. J. — Infantry Weapons of the World (1977)
Gordon-Douglas, S. R. — German Combat Uniforms 1939-1945 (1970)
Gregory, Barry — British Airborne Troops (1974)
Hogg, Ian — German Artillery of World War Two (1975)
Huston, James A. — Out of the Blue; U.S. Army Airborne Operations WW2 (1973)
Low, Professor A. M. — Parachutes in Peace and War (1942)
Millar, George — The Bruneval Raid — Flashpoint of the Radar War (1974)
Mollo, A. and McGregor, Malcolm — Army Uniforms of World War 2 (1973)
Mrazek, James E. — The Fall of Eben Emael — Prelude to Dunkirk (1970)
Ryan, Cornelius — A Bridge Too Far (1974)
Stahl, Peter — Fallschirmjager 1936-1945 (1969)
Tugwell, Maurice — Airborne to Battle — A History of Airborne Warfare 1918-1971 (1971)
Weeks, John — Airborne Equipment — A History of its Development (1976)
Wise, T. — American Military Camouflage and Markings 1939-45 (1973)

Key Uniform Guides

1 — United States Infantry Europe 1942-45 by Howard P. Davies (1974)
2 — British Parachute Forces 1940-45 by Howard P. Davies (1974)
3 — U.S. Airborne Forces Europe 1942-45 by Brian L. Davis (1974)

World War 1 Fact Files

Axis Pistols, Rifles and Grenades by Peter Chamberlain and Terry Gander (1976) Infantry, Mountain and Airborne Guns by Peter Chamberlain and Terry Gander (1975)
Allied Pistols, Rifles and Grenades by Peter Chamberlain and Terry Gander (1976)
Sub-Machine Guns and Automatic Rifles by Peter Chamberlain and Terry Gander (1976)
World War II Special — Airborne Invasions — Paratroops and Glider-borne Forces in Action (1976)

The best known books in the fast-growing list of wargames literature are —

Little Wars, H. G. Wells (1913)
Wargames, D. F. Featherstone (1962)
Wargames in Miniature, J. Morschauser (1963)
Naval Wargames, D. F. Featherstone (1966)
Air Wargames, D. F. Featherstone (1967).
Charge!, Brigadier P. Young and Lt-Col. J. P. Lawford (1967)
Advanced Wargames, D. F. Featherstone (1969)
Discovering Wargaming, John Tunstill (1969)
Introduction to Battle Gaming, Terence Wise (1969)
Battle! Practical Wargaming (World War II), Charles Grant (1970)
Battles with Model Soldiers, D. F. Featherstone (1970)

Wargames Campaigns, D. F. Featherstone (1970)
The Wargame, Charles Grant (1971)
Solo Wargames, D. F. Featherstone (1972)
Wargames Through the Ages, Vol. I Ancient and Medieval Periods, D. F. Featherstone (1973)
Battle Notes for Wargamers, D. F. Featherstone (1973)
Tank Battles in Miniature — The Western Desert Campaign, D. F. Featherstone (1973)
Napoleonic Wargaming, Charles Grant (1973)
Wargames Through the Ages, Vol. II 1420-1783, D. F. Featherstone (1974)
Practical Wargaming, C. F. Wesencraft (1974)
Battle Notes for Wargamers — Ancient and Medieval Periods, D. F. Featherstone (1974)
An Introduction to Wargaming, John Sandars (1975)
Tank Battles in Miniature: The Russian Campaign, Bruce Quarrie (1975)
Wargames Through The Ages 1792-1859, D. F. Featherstone (1975)
Wargames Through The Ages 1792-1859 Vol. III, D. Featherstone (1975)
Tank Battles in Miniature: North-West European Campaign, Bruce Quarrie (1976)
World War Two Wargaming, Bruce Quarrie (1976)
A First Book of Wargaming, F. E. Perry (1976)
Operation Warboard, Gavin Lyall (1976)
Wargaming Pike and Shot, D. F. Featherstone (1977)
Wargames Through The Ages: Vol. IV 1861-1945, D. Featherstone (1976)
Wargamers Handbook of the American War of Independence, D. Featherstone (1977)